THE MATH PLAGUE

HOW TO SURVIVE SCHOOL MATHEMATICS

Sherry Mantyka

MayT Consulting Corporation
St. John's, Newfoundland, Canada

2007

08 07 06 05 04 03 02 01 1 2 3 4 5 6 7

Library and Archives Canada Cataloguing in Publication

Mantyka, Sherry, 1949-
   The math plague : how to survive school mathematics /
Sherry Mantyka.

Includes bibliographical references.
ISBN-13: 978-0-9781658-0-2
ISBN-10: 0-9781658-0-2

   1. Mathematics--Study and teaching. I. Title.

QA40.M36 2007   510'.71   C2006-905989-6

MayT Consulting Corporation
84 Allandale Road
St John's, Newfoundland, Canada   A1B 3A1

Printed in Canada

TABLE OF CONTENTS

PREFACE

I am writing this book for students, teachers and parents who are frustrated with the current state of mathematics education. This work has been inspired by the countless inquiries I have received as Director of the Mathematics Learning Centre. The content of this book is the amalgamation of what I have learned from all the courageous individuals I have worked with for the past eighteen years who have honored me with their trust. It is to all these individuals that I dedicate this work.

The book is divided into thirty-eight small sections, each containing a confirmed principle for the effective learning of mathematics. Each section is introduced with a quotation, most of which are by well-known public figures. The relevance of the quotation to the foundational principle or practice for the learning of mathematics is explained, and then more detail about implementing the principle is provided.

This book is applicable to students of mathematics of all ages. Whether or not the book is read in its entirety, the reader will relate to situations described and find concrete suggestions for enhancing the learning experience in those situations.

If any of the principles and practices are adopted, success in the learning of useful mathematics will improve. This does not happen by trying to make the learning of mathematics more enjoyable or easier. Rather it happens by increasing the learner's willingness to focus on the difficult material, and therefore making whatever time a student spends studying mathematics as productive as possible.

ACKNOWLEDGEMENTS

I would like to thank Dr. Peter Schotch, Department of Philosophy, Dalhousie University, for his thoughtful and extensive suggestions for revisions, almost all of which I wholeheartedly embraced.

I am also grateful to the full-time staff at the Mathematics Learning Centre (Selena Delahunty, Sheila Penton, Theresa Kelly, Geraldine Walsh and Christine McDonald) who painstakingly edited and compiled the bibliography and footnote references for the entire manuscript.

I am grateful to my brother, Don, for his indulgence in engaging me in the various thought experiments and somewhat radical joint presentations we did. I am also indebted to Don for writing a section of this book on linebacker training and for editing other sections. I am also grateful to my sister-in-law, Arleigh, for putting up with conversations about these things in her living room.

I must also thank David Kelleher-Flight for his willingness to explain in great detail how he has trained to be a vocalist, and also, how he achieved scholarship level grades in mathematics in spite of his distaste for the subject.

Lastly, there are no words to adequately express my gratitude to my secretary, Denise Porter, who formatted the entire manuscript and also made it more of a "work of art" and less of a mere sequence of paragraphs.

INTRODUCTION

I have called this book *The Math Plague* because that is exactly how under-achievers in mathematics feel about the subject. It is neither enjoyable nor gratifying in any way. Rather it is an unavoidably frustrating and disagreeable experience, and with its torments and vexations can be likened to the black plague or a plague of insects. Figuratively and emotionally, this is not an exaggerated metaphor. Being forced to do school mathematics year after year with little success and seemingly little hope for a brighter experience in subsequent grades, can be an almost soul-destroying experience for many. Indeed for these individuals, who comprise about 80% of the population, mathematics and its pervasiveness in today's world accounts for many educational drop-outs.

In the affluent post World War II era in North America, it has been thought that education is the panacea to all social problems. This has meant that social policies have been structured as much as possible to keep youth in the educational stream as long as possible. Making more education financially available to students was one aspect of this social agenda that had to be addressed, but having students successfully complete academic programs was also a huge problem. It is the latter of these two issues that I will address in this book.

There are basically two ways to achieve higher completion rates in educational programs. The easiest way is to lower academic standards. The hardest way is to seek an understanding of why students under-achieve and then develop teaching/learning strategies to address these issues.

There is also a potentially dangerous middle-ground which provides new, alternate educational programs. This is a sound strategy if it is based on the latter premise above; it is not a sound strategy if it is merely a lower standard academic program with a different name. Academic standards that have survived the test of time have done so for a reason – they are based on providing an educational experience that turns out to be useful. Formal educational attainment that does not provide individuals with any ability to be more productive in the work force or more enriched in their personal lives is a waste of everybody's time and money. Illusions such as these belong only in amusement parks and on the entertainment stage, not in our educational system.

The principles and practices outlined in this book form the basis for our operation at the Mathematics Learning Centre. Our Centre primarily serves students who, upon entering university, fail the mathematics placement test. Those students who complete our program not only overcome their academically disadvantaged position in university mathematics courses, but also university English courses. Furthermore their graduation rate from university doubles over those who fail the placement test and do not complete our program.

Paramount to my work at the Mathematics Learning Centre in developing and employing teaching/learning strategies, has been my involvement with athletic coaches. This is for two reasons. Firstly, there are many analogies between learning to play a sport well and learning to do mathematics well. Secondly, coaching clinics tend to be far more inspirational to coaches than teacher in-services are to teachers. My brother, Don, is both an inspirational high school physics teacher and football and curling coach. We have spent many hours talking about the relationship between the two. We have done joint presentations at academic gatherings about these issues in mathematics education. We now firmly believe that for teacher in-services to be effective, they must be conducted in a manner similar to coaching clinics – there must be an appeal to the passion which lies within us all. And then as educators we must reflect this passion to our students when we work with them.

"Seventeen-year old Czech tennis sensation Nicole Vaidisova, a semifinalist at the French Open and high school senior, when asked if she is nervous about playing at Wimbledon: 'Math is what worries me most,' she told reporters of her upcoming final exams. 'That is always a struggle.' "

(Hutchinson, 2006)

"Because of its climate and geography, Newfoundland is ideally suited for the production of alcoholics, royal commissions, snow, unsolvable enigmas, self-pity, mosquitoes and black flies, inferiority complexes, delusions of grandeur, savage irony, impotent malice, unwarranted optimism, entirely justified despair . . ."

- Quote by Wayne Johnston from his book, *Baltimore's Mansion* (1999, p. 123)

When I was in Australia writing this book, I gave a talk about my work to Newcastle Enterprising Women. The speaker that I followed did a wonderful presentation about his time living with monks in Tibetan monasteries. I had fully prepared my talk before I heard him speak, but as I listened to him, I decided that I needed to completely re-vamp my introductory remarks to follow his. I chose the quotation on the left-hand page, accounting for my choice by saying, "How could you possibly believe that anyone from my home territory, Newfoundland, would have anything of use to say to you?". Then, somewhere in the middle of my presentation, I asked my audience to write on a piece of paper that they would submit to me to keep, whether they perceived there to be a math problem in Australia, and if so, what that problem was. There was one Canadian in the audience who had married a handsome Australian and had been living in Australia for twenty years. Knowing that I had been raised in a different province of Canada, Saskatchewan, she wrote on her slip of paper, "Would you say your sense of humor has been influenced since you went to Newfoundland?" My answer was a simple and emphatic yes.

I often use Wayne Johnston's quotation in my presentations to policy committees to account for our excellence, our leadership, our forthrightness in tackling the math problem head on. Newfoundland is a large island with a small population and an historically resource-based economy. We are struggling to find our place in an intensely technological world economy. We have too few people to be able to disregard any of them as participants in this new economy. We must train all of our youth to be as skillful as they can be in this new economic milieu and that means they must have strong mathematical skills. Calculators and computers do NOT replace key mental processes. Managing limited working memory is a key mental process. Mathematics is a subject well suited to learning how to manage limited working memory, and a facility to mentally manipulate numbers and symbols is fundamental to being able to function in a technological world.

This is why we are in such a strong position to provide leadership in assisting the mathematically uninspired to become competent mathematical practitioners. It is because, as Johnston so eloquently puts it, Newfoundland embodies all extremes.

"Once I get a good look, it's all mechanics from that point on."

- Quote from Michael Jordan of the *Wizards* in an interview after scoring the game winning point against the *Cavaliers* on January 31, 2002 (TSN SPORTSCENTER)

In athletics, it is taken for granted that excellence entails hours and hours of repetitious drills of basic skills. In basketball, the basic skills include dribbling, lay-ups, free throws. Practicing these skills individually in lengthy sessions of boring, repetitious activities allows the brain to code the body to respond automatically when execution is required in a game situation. This frees the brain's working memory to dedicate its processing to strategic implementation of the various skills required to play an excellent game of basketball. This is what Michael Jordan is talking about. Once his brain's working memory has identified a scoring opportunity and a strategic plan for implementation, his body will take over. The well-trained elite athlete will not have to think about the specific dynamics of how to successfully dribble the ball to the right point on the basketball court to score the basket. "It's all mechanics from that point on." (TSN SPORTSCENTER, 2002)

The same is true about using mathematics in problem solving. If you have practiced all the basic skills needed to solve problems to the point of automaticity, then you will not have to use any working memory to remember multiplication facts, how to factor polynomials, or laws of exponents – all the basic elements of useful mathematics. Instead, you will be able to devote all your working memory capacity to the more demanding task of figuring out strategies for solving the problem. If you leave it to your working memory to try to figure out what skills are needed and how to employ those skills, then you will overload your brain and make mistakes.

"Recognize that new habits need practice, practice, practice until they become your own."

- Taken from *Tools for Living, Winning Points, Weight Watchers,* one of nine tips "to help make those diet resolutions a life-long achievement"
("Weight Watchers Offer New Year's Diet Tips," 2002)

It is interesting to find the word "practice" in the context of eating patterns, and to notice the repetitious use of this word in the quotation from Weight Watchers ("Weight Watchers Offer New Year's Diet Tips," 2002). It suggests that what works well in one aspect of life-style management may indeed work well in others, and that it is hard to avoid the tedium that comes with having to do the same thing many times over. Practice and boring repetition are essential to success.

Becoming a truly successful student of mathematics entails more than just taking a course and getting a good grade. A successful student of mathematics is open to a life-long learning experience of mathematics. This student does not engage in terminal mathematics courses. What we are talking about here is a fundamental change in attitude towards the learning of mathematics, a change in attitude which is similar to that described in the Weight Watchers brochure: "Levels of Change - Losing weight and keeping it off involves change – change in how one thinks, feels and behaves" ("Weight Watchers Offer New Year's Diet Tips," 2002).

Becoming a successful student of mathematics will entail being attentive in classes, choosing to spend more time practicing basic skills, persisting with learning the parts of mathematics that you find difficult, trusting your teachers, and believing that it is within you to learn how to use mathematics when you need to.

"Perhaps it was because he was now so busy, what with Quidditch practice three evenings a week on top of all his homework, but Harry could hardly believe it when he realized that he'd already been at Hogwarts two months. The castle felt more like home than Privet Drive had ever done. His lessons, too, were becoming more and more interesting now that they had mastered the basics."

- Taken from Harry Potter and the Philosopher's Stone
(Rowling, 2000, p. 126)

If a gifted wizard like Harry Potter is expected to master the basics of magic before tackling complex spells like transfiguration, then surely we mere Muggles must expect to have to master the basics of mathematics before tackling complex problem-solving.

This is not meant to undermine the ultimate goal at all; rather it is meant to facilitate its achievement. Even though the focus of the upgrading program at the Mathematics Learning Centre seems to be skills acquisition, it is really the ability to be an independent learner that is of paramount importance. This objective embodies the acquisition of study skills, analytical skills, and the ability to reason and problem-solve.

Problem solving is included in all modules of our program. Almost every module test has a section containing only word problems for which the acceptable pass standard (with no part marks awarded) is between 50 and 65 percent.

In addition, students are constantly reminded of the role of pattern recognition in numeric and symbolic manipulation. Even with simple arithmetic problems (e.g., 0.88/0.044), students are reminded that the easiest way to answer these questions is to recognize and make use of place value manipulations and the cancellation of common factors. This means that students cannot be permitted the use of a calculator during the early stages of the arithmetic portion of their personal programs. Once they have built up or restored their facility with basic number facts and place value computational skills, they tend to use these methods even when a calculator is handy. More importantly, they develop a sensitivity to numerical patterns which can be transferred to algebra.

After students have been rescued from their total dependency on calculators, they are exposed to their actual usefulness and taught how to fully exploit their power in the execution of computations. For example, students are taught how to use the calculator memory and all the calculator functions to facilitate internal storage of the maximum number of digits. They are also taught the appropriate rounding rules and rules associated with the propagation of measurement errors in calculations. Finally, they are expected to be able to put all these elements together in solving context-based problems entailing measurement and computation in various settings, including trigonometry.

Just as for Harry Potter, mastering the basics is a means to an end. Whether it be Muggle mathematics or Wizard magic, the basics must come first.

"Math teachers are badly dressed, overweight, scruffy and friendless, according to students in North America and Europe. Researchers at Plymouth University in England talked to 12- and 13-year old pupils in Britain, Sweden, Norway, Finland, Germany, Romania and the United States about their perceptions of math teachers and found the results were overwhelmingly negative, reports the BBC. Other features associated with math teachers were beards (they were almost always seen as men), baldness, bad haircuts, holes in their clothes and a bleak social life. Said Professor John Berry: 'One worrying aspect is that children may be put off studying math if they think others will see them as being nerds.' "

- Taken from *The Globe and Mail*,
January 22, 2001
("In the News," 2001, p. A16)

When I was a young woman doing my Honors degree in Mathematics, the issues raised in the excerpt on the left-hand page were of grave concern to me. All my professors broadly fit that description. Furthermore I wanted to go on to do a Ph.D. in Mathematics, but I was also a semi-professional dancer at the time and I dressed very fashionably. I wondered how on earth I would ever survive in a profession if my peer group could be accurately described in that manner.

The answer: socialize outside your professional peers. It takes a little more effort but it can be done and I did it. My best friend in high school and university is a very successful visual artist, and my social circles have always tended to be centered in the arts community. I continue to enjoy fashions, I dress fashionably, and I have a lively social life. I also have never worn a beard nor do I wear clothes with holes in them (although my first-year calculus professor did come to class one day with a huge hole in his trousers that he had patched up with scotch tape). In fact, I have thoroughly enjoyed my involvement in my profession, bringing the non-traditional elements into play there. If you are so inclined, I encourage you to do so also – it can be fun raising eyebrows and challenging stereotypes.

"We Learn By Doing

Not many years ago I began to play the cello. Most people would say that what I am doing is 'learning to play' the cello. But these words carry into our minds the strange idea that there exists two very different processes: (1) learning to play the cello; and (2) playing the cello. They imply that I will do the first until I have completed it, at which point I will stop the first process and begin the second. In short, I will go on 'learning to play' until I have 'learned to play' and then I will begin to play. Of course, this is nonsense. There are not two processes, but one. We learn to do something by doing it. There is no other way."

- John Holt, taken from
Chicken Soup for the Soul
(Canfield and Hansen, 1993, p. 129)

"Cognitive psychology is the branch of psychology concerned with the functioning of the brain, especially in regards to mental processes operating in response to stimuli, such as how well a student learns in response to different teaching methods" (Collins and Harding, 2004, p. 9). In the literature of cognitive psychology, the "learning by doing" process described in the quotation on the left-hand page is called *implicit learning*. This is the way most humans learn their first language. This is also the way most humans learn to play a sport or a musical instrument, or learn how to dance. However, contrary to Mr. Holt's contention in his quote, it is not the only way humans learn. There is another process called *explicit learning*, and it is this process which is more common and more effective in the learning of mathematics.

Implicit learning is characterized by imitating behavior without necessarily acknowledging the underlying rules that determine this behavior. For example, children learn to speak their first language employing correct rules of grammar without being able to name those rules of grammar. On the other hand, explicit learning is rules oriented. Rules are learned, practiced in drills, and then implemented in problem situations. This is the most common way that adults learn a second language. They memorize the alphabet of the new language, they learn some vocabulary, they learn the rudimentary rules of grammar, and then they attempt to string words together according to the rules of grammar to generate simple sentences. With enough practice, this leads to an ability to converse in that language.

The rules-oriented approach to the teaching of mathematics is immediately recognizable by all. As children we are taught to memorize addition facts and then multiplication facts. Then we are taught algorithms for using these facts, and simple but abstract relationships between addition and subtraction and multiplication and division to enable us to subtract and divide whole numbers. In junior high school we are taught to employ these facts and these rules in more complex procedures for dealing with all the analogous arithmetic operations for ratios of whole numbers, fractions. Facts to be memorized, rules to be learned, practice to merge the two until they both become second-nature, then more complicated rules to be learned which build on the simple ones, and more practice to merge it all until it becomes second-nature, and on and on and on it goes. In fact, it goes on beyond what most naturally inclined implicit learners can tolerate after the age of twelve. That is when most implicit learners stop acquiring new knowledge of mathematics.

If this is so, why can't mathematics be taught to those individuals implicitly? To a point, it can. In fact, once rules of mathematics have become second-nature as described in the previous paragraph, then invocation of those rules is more in keeping with the process of implicit learning. That is, correct rules are employed without explicit acknowledgement of that rule at that moment. At this ideal stage of learning mathematics, this process is really a combination of the two. Unfortunately, without a lot of drill and practice of all the basic skills in mathematics, the learner will not achieve this ideal.

Another element of implicit learning that makes it ill-suited to the learning of mathematics is the lack of *transference*. In general, implicit learning is associative and context bound, while explicit learning is based on hypothesis testing and transfers across contexts (Berry & Dienes, 1993). Many basic mathematics skills can be learned associatively, but skills learned this way will be context bound and of limited value without an explicit understanding. For example, a child who cannot describe the conceptual meaning of multiplication but is merely drilled on multiplication tables, has in effect learned multiplication facts associatively. As a result, this child might be able to quickly state the answer to 7×9 but would fail to use multiplication when asked about the number of cups in a 7×9 array.

If the mathematical skills we learn are to be useful across different contexts without learning and practicing them in all potentially relevant contexts, then it is essential that learning be explicit. Automaticity of a skill is about making an explicitly learned skill also available implicitly. Thus, in the context or contexts in which the skill is highly over-learned (that is, is second-nature) it will be appropriately and quickly retrieved with limited investment of mental resources. However, when required, the skill will be explicitly accessible and, therefore, be useful in new contexts. Driving an automobile is a skill most of us acquire explicitly which eventually becomes automatic. This skill is ordinarily executed with so little demand on resources that most of us fail to remember what we did and what we saw on the way to work. On the other hand, we can explicitly retrieve and describe the relevant sub-skills when teaching someone to drive. Furthermore, we do explicitly use and monitor these skills when road conditions are poor.

This is the ideal for mathematical skills to be useful in problem solving. When routine situations are encountered, we want invocation of mathematical procedures to be automatic so that working memory can attend to higher order thinking. On the other hand, when unfamiliar problem situations in mathematics confront us, we want to be able to retrieve explicit rules and use our higher order thinking to creatively manipulate those rules to fit the new context (May, Rabinowitz and Mantyka, 2002).

"Perfection happens only when the mechanics are automatic."

- *The Little Book of Coaching*, written by
Ken Blanchard and Don Shula
(the winningest coach in NFL history)
(2001, p. 44)

Don Shula is a legendary football figure in the United States of America. He led the Miami Dolphins to five Super Bowl appearances and the Baltimore Colts to one. He is the only NFL coach to have a perfect 17-0 season.

We have already drawn many analogies between athletics and doing mathematics. Here is another one – the importance of being perfect, and how to attain that perfection.

Mathematics is meant to be useful. Answers which are 70% correct are not useful. Therefore to be useful, mathematics must be perfect, 100% correct. There must be no copying errors, no careless multiplication fact errors, no un-labeled axes, no missing units in word problems.

It is the same in football. If only one of five offensive linemen misses his block, the running back will probably run into a brick wall of defensive players and gain little or no yards on the play. So a play which is only 80% correct in execution is unlikely to be of any use to the team.

What creates perfection in execution? Shula maintains it is automaticity in the execution of mechanics. He does not want his players to be worried about what they should be doing in a game situation because they will hold back. He wants them to be executing their assignments automatically, on "autopilot," as he puts it. (Blanchard and Shula, 2001, p. 45)

Shula's belief is supported in the literature of cognitive psychology which we have already cited. Automatic responses to low-level skills (the mechanics) free the working memory to focus on achieving the higher levels of performance which Shula demands from his players.

Learning to do mathematics is not unlike learning how to play football. To be a football player who can meet Shula's winning standards, one must spend hours in repetitious practice of the mechanics. To be good at mathematics, one must spend hours in repetitious practice of the mechanics – basic skills, like multiplication facts, laws of exponents, and factoring polynomials. One can participate recreationally without this level of commitment, but recreational involvement does not lead to further skills development. If one's involvement with learning is designed to open doors and facilitate better opportunities for a career or for personal development, then a recreational involvement with learning simply will not suffice. One must aspire to perfection and one must be willing to achieve this perfection through repetitious practice of the mechanics until automaticity is achieved. Only then can excellence be achieved.

"Some guys think they'll do what they need to do when it really matters, but it always matters. You see, you're telling yourself and your teammates who you are every day – when you don't play defense or pass to an open teammate, when you show up late or don't hustle, or when you seek too much credit or refuse to accept criticism. With every action, you're revealing your character. It's impossible to be responsible occasionally."

- Pete Carril, 71, legendary basketball coach at Princeton and in the NBA, taken from Where Pride Still Matters, *Men's Health* (Brooks, 2002, pp. 80-83)

Aspiring to excellence is not a sometime thing; it is a matter of habit. It is not okay to attend classes most of the time. It is not good enough to practice mathematics only when you are in the mood to do so.

A familiarity with mathematics is nice. Understanding mathematical concepts is fine. But to be able to do useful mathematics, a mere familiarity and/or understanding will not suffice. Being able to do mathematics comes with hours of regular practice, much of which is repetitive. Furthermore, if the lags between practice sessions are too long, the skills will not be retained.

If you are going to do mathematics, then you must accept the fact that your commitment to the subject is long-term. You cannot simply study the few or many courses that are required for your formal education and then forget it all. As Pete Carril puts it in the quotation on the left-hand page, "It's impossible to be responsible occasionally."

Drew Bledsoe, former $100 million quarterback for the New England Patriots, understands this principle all too well. In the 2001-02 football season, he was injured early in the season and replaced by a young unseasoned quarterback, Tom Brady. Quite unexpectedly the team did very well under Tom's leadership at quarterback so that when Drew Bledsoe was able to return to the team to play, the coach chose Tom Brady as the starting quarterback. Although Drew Bledsoe desperately wanted to be playing, he stoically accepted the coach's decision. Then, in the AFC Championship game, Tom Brady got injured in the second quarter and Bledsoe, having not thrown a football in the heat of competition for four months, was back in the game. The following quotations are taken from the article "A dream sequence for player who never quit" (McDonough, 2003, p, D8), describing Bledsoe's courageous performance that helped the Patriots to a 24-17 win over the Steelers and earned them a berth in Super Bowl XXXVI:

I thought about this moment for months . . . . I've had the same dream for weeks . . . . I saw myself in the game. I knew what the plays were. I played them out in my dreams. Where to throw the ball, what to look for. I had all kinds of game situations. In my dreams I saw the touchdown pass I threw today. That's why I was excited when the call came in. That's one of my favorite pass routes. The deep corner. I know I can throw that one. And then it came. I saw it developing and I knew what I had to do. It's one thing to think about these situations, and dream about them, it's another thing to go out and do it. You can't freeze when you have the chance. You must be just as aggressive as you were when you played it through in your mind. That's what I told myself today. Don't back off. Make the tough throws. Go after these guys. And I did. I wasn't afraid. I wanted to stay aggressive and we did.

That's exactly what Pete Carril is talking about in his quotation, and that's exactly what you have to be willing to do if you aspire to excellence in whatever you do – practice until you drop, deal with setbacks, accept disappointments without giving up. It's more a matter of character than of content.

"**Character cannot be developed in peace and quiet. Only through experience of trial and suffering can the soul be strengthened, ambition inspired, and success achieved.**"

- Taken from *Curves® Member Guide*
(Heavin, Findley and Thomas, 2005, p. 115)

"Hard work. Practice. Repetition. Belief in the system. And a will to prove something. These are the blocks on which the 2001-02 Patriots are built."

- From "AFC Championship Patriots 24, Steelers 17," *The Boston Globe*, January 28, 2002, p. A1

The 2001-02 football season was indeed a memorable one for the New England Patriots football team and its fans. The New England team had only made it to the Super Bowl twice previously and had lost both times by a cumulative score of 81-31. Furthermore, the team's last trip to the Super Bowl had been sixteen years prior.

The season leading up to the AFC Championship title had been a tough one for the New England Patriots. Their $100 million quarterback had been injured early in the season and was not able to return to the game until very late in the season. In the interim a young quarterback had taken over the position and did exceptionally well. The team had won nine games in a row leading up to the AFC championship game under the young quarterback's direction at this key position. So the coaching staff had this dilemma to face – who would be the starting quarterback in the AFC championship game?

The New England Patriots were the underdog team for most of that season. As they continually defied the odds with consecutive wins following their quarterback's injury, they were tagged the Cinderella team, the team of destiny. The coaching staff was masterful in managing the team's situation. The head coach, Bill Belichuk, was heard repeatedly in interviews following winning games to speak of his respect for his players' commitment to the team and to hard work and repetitious practice of what would turn out to be key plays in many tough game situations that season. The team's manner of dealing with their situation that season became a model for dealing with tough situations in general and indeed deserved to be declared a Cinderella team, a team of destiny.

The analogy with learning mathematics is simple. You may be someone who does not enjoy the drill and practice associated with learning mathematics. You may not recognize how mathematics will be of any use to you in what you want to do with your life. You may find it takes twice as much practice for you to develop the same level of mathematical competence as your friends. You are a mathematics underdog – you do not have a $100 million dollar quarterback in your head guiding your learning of mathematics. But something in your environment is telling you your season is far from over. So what do you do about it?

You do have choices here. You can allow yourself to give in to the despair that comes with knowing your future is not going to be easy. Or you can face the challenge with honor and dignity and self-respect. You can choose to make the most of every moment without being overly concerned with the final outcome. You can choose to trust those who are trying to guide you through this tough experience. You can choose to work hard at distasteful tasks because you believe in your ability to overcome these obstacles. You can choose to be a Cinderella and not an evil stepsister. You can choose to be a person of destiny and not a loser.

I have worked with students who find doing mathematics easy and I have worked with students who find doing mathematics really difficult. The students I have the most respect for are the ones who work hard at it no matter what. There is less achieved by someone who finds the task easy than is achieved by someone who overcomes great obstacles in reaching the same goal. The key element is to be a person of strong character – someone who wants to make the most of your life, no matter what the circumstances are.

"His ragged little group of veterans, utterly without honor in their previous armies, were blossoming into capable leaders . . . . He had the army drill in eight-man [pla]toon maneuvers and four-man half-toons, so that at a single command, his army could be assigned as many as ten separate maneuvers and carry them out at once. No army had ever fragmented itself like that before, but Ender was not planning to do anything that had been done before, either. Most armies practiced mass maneuvers, preformed strategies. Ender had none. Instead he trained his toon leaders to use their small units effectively in achieving limited goals. Unsupported, alone, on their own initiative. He staged mock wars after the first week, savage affairs in the practice room that left everybody exhausted. But he knew, with less than a month of training, that his army had the potential of being the best fighting group ever to play the game."

- Taken from *Ender's Game* by
Orson Scott Card (1994, p. 175)

"When Belichuk
was asked about a blocked
field goal that resulted
in a Patriots touchdown
(including a nifty lateral), he
said 'That's another one of
those drills we work on every
single Thursday in practice
and we've done it a thousand
times. I can't tell you how
many times that play has
happened in practice, and
to see it work the way it did
out on the field today, that's
where all of that hard work
pays off.' "

- Excerpt from "Next Stop, Super Bowl,"
The Boston Globe,
January 28, 2002, p. D9

Developments in cognitive psychology and intervention research provide the rationale for our insistence on the importance of repetitious drill and practice in the learning of mathematics, just as Coach Belichuk acknowledges it in his comments to the newspaper reporter regarding his team's success in the AFC Championship game.

Over the past two decades, cognitive psychologists have proposed a variety of models initially designed to account for errors that occur when people are required to perform two or more tasks simultaneously. In most of these models, it is assumed that some resource (e.g., energy, attention, working memory) is limited. Three ideas which are relevant to mathematics are: (a) capacity limits performance – mathematical performance will eventually break down as task difficulty is increased; (b) capacity is used in performance – adding more components to a problem (e.g., presenting a word problem rather than simply stating the problem numerically) will use more capacity and increase task difficulty; and (c) demands on capacity are less for individuals who have higher levels of relevant skills.

This is why we insist that students of mathematics must be prepared to dedicate much of their time learning mathematics to the practice of basic skills in order to be good problem solvers. From the perspective of this research in cognitive psychology, an individual's performance in problem-solving will be maximized if they do. The more attention that has to be given to execution of specific skills while a problem is being solved, the more mistakes will be made.

This principle is taken for granted in sports but is ignored in modern mathematics school curricula. Coach Belichuk knew that the only way for his team to be able to successfully execute a blocked field goal in a game situation was to practice it over and over and over again in non-game situations. It is the same brain driving our bodies when we play a sport as it is when we try to solve a problem using mathematics. Why on earth would we expect to be able to circumvent the drill and practice in mathematics when we know we cannot circumvent it in athletics?

"CHAMPIONS!
Patriots win Super Bowl, 20-17, on Vinatieri's last-second kick

These Patriots seemed to thrive in their position as underdogs. A last-place team one year ago, listed at 75-1 to win the Super Bowl when the season started, they jelled under the leadership of coach Bill Belichick and the quarterbacking of Brady. They won their final nine games, and last night's shocker ranks as the second-greatest upset in Super Bowl history . . ."

- Taken from *The Boston Globe*
(Shaughnessy, 2002, p. A1)

For students who are mathematically disadvantaged, intervention programs are required to improve the situation. Schorr (1989) provided an overview of intervention programs designed to improve the lives of disadvantaged children and their families. Two of her conclusions were: (a) bad outcomes are always a consequence of multiple risk factors; and (b) effective programs, which are always comprehensive, change predicted lifetime trajectories. We believe both conclusions are relevant to conceptions and practices involving remedying sub-standard performance in mathematics. First, it is unlikely that there is a single causative factor which produces such extreme mathematical disability. If so, there will be no simple fix at the elementary, junior high school, or high school level that magically eliminates mathematical deficiencies. Second, it is likely that only well-conceived, comprehensive catch-up programs which focus on both mathematical deficiencies and cognitive/emotional problems will be effective.

Schorr (1989) also abstracted a number of features that are common in effective intervention programs, some of which are essential to incorporate into intervention programs for mathematically challenged students. Those features include: (a) a broad spectrum of services (i.e., teaching methods, tools, listening to students, and referral to counseling when appropriate); (b) flexible staff members and program structures (i.e., the program should be structured and restructured according to the current needs of each student); (c) mutual respect and trust between student and staff; and (d) services that are coherent and easy to use. Schorr reported that three of the consequences of effective educational interventions with preschoolers and school-aged children that are reported by both children and their parent(s) are: empowerment (i.e., a belief that what the student does can make a difference in outcomes); change in attitude (i.e., a belief that education is useful); and expectation (i.e., a belief that a student can achieve goals originally thought to be unlikely or impossible).

It is at this point that we see the connection to the 2002 Super Bowl victory of the Patriots.

The coaching staff of the Patriots provided training and an environment for that training that was fully cognizant of the team's position in the league and the individual strengths and weaknesses of its players. This environment fostered a belief that they could succeed as underdogs if the team gave its full attention to each moment of practice, to each moment of the games, without regard for final outcomes. The coach respected the players as individuals and believed in their ability and willingness to work hard towards a common goal. The players trusted the coaching staff's leadership and committed themselves fully to the drills in practices and to tough, physical game situations.

Coach Belichick had a difficult decision to make regarding which quarterback would start in the Super Bowl game – the $100 million Bledsoe, or Tom Brady, who would, as a result of Belichick's decision, become the youngest quarterback to win a Super Bowl game. In the week preceding, both quarterbacks when asked by reporters whether or not they would be starting, replied simply, "You better ask Belichick" (MacMillan, 2002, p. D8). In a newspaper article in *The Boston Globe* entitled "Bledsoe's comeback is a real tear-jerker," Michael Holley reports: "For a few moments, Bledsoe vs. Brady was insignificant. If you watch the way Bledsoe and Brady interact, you understand why the Patriots seem to play better than anyone expects. They're a team. Just like their quarterbacks." (2002, pp. D1, D8)

Mutual trust and respect - it's a winning combination in athletics and it's a winning combination in the classroom.

"Ericsson and his colleagues have thus taken to studying expert performers in a wide range of pursuits, including soccer, golf, surgery, piano playing, Scrabble, writing, chess, software design, stock picking and darts . . . . Their work, compiled in the "Cambridge Handbook of Expertise and Expert Performance," a 900-page academic book that will be published next month, makes a rather startling assertion: The trait we commonly call talent is highly overrated. Or, put another way, expert performers – whether in memory or surgery, ballet or computer programming – are nearly always made, not born. And yes, practice does make perfect."

(Dubner and Levitt, 2006)

[Stephen J. Dubner and Steven D. Levitt are the authors of "Freakonomics: A Rogue Economist Explores the Hidden Side of Everything." More information on the research behind this column is at www.freakonomics.com.]

"The real voyage of discovery consists not in seeking new lands but in seeing with new eyes."

- Marcel Proust
(Scattergood, 1884)

As someone who has had the opportunity through her work to travel to many new lands, I found the quotation on the left-hand page to be particularly provocative. I naturally wondered if I had been deluding myself into thinking I had been taking voyages of discovery when in fact I was only seeing new things with old eyes.

In this section I have constructed a special voyage of discovery for the reader. I asked my conference co-presenter, Don Mantyka, to write a description of the *triggers* that a defensive linebacker is trained to use in football (grid-iron in Australia). In so doing, Don actually constructed a complete description of the training that entails use of triggers (*memory retrieval*), repetitious practice to obtain automaticity, and working memory processing of game situations (*deliberate attentional resources, supervisory attentional mechanisms*). As I was reading Don's description, I wasn't surprised to see direct analogies to the teaching/learning of mathematics. But what was unexpected was recognizing it as a description of the *whole* process and the final desirable outcome.

What follows is in three parts. Part I is Don's description of a linebacker's training. Part II is a matching quiz. The sections to be matched from Part I are enclosed by square brackets with a letter in square brackets following. The descriptions of mathematics are numbered. There is one number which is a best match for each letter. Part III highlights the connections suggested in the matching quiz.

This challenge for the reader is the reader's personal voyage of discovery. For those readers who are used to having answers at the back of mathematics textbooks, answers can be found on my website (Mantyka, 2006).

Part I.

A linebacker in football relies on a combination of instinct and conditioned responses.

The player has to start each play in a continually practiced stance. At the instant that the ball is put into play the linebacker is conditioned to take a jab step with his inside foot. This physical movement is a *switch* to start the cognitive analysis of the developing play.

The first *key* to be read is the offensive lineman directly opposite the linebacker. [If the lineman moves forward, the linebacker is taught to meet him head-on. If the offensive man backs up in a pass block, the linebacker retreats to his pass coverage responsibilities.][A] The coaching phrase is, "he comes, I come; he drops, I drop." This is drilled on a regular basis, often with two players just going through the first two steps of the play. [If the offensive player moves laterally down the line of scrimmage going to throw a block for a play being run to the outside, the linebacker is coached to follow him down the line always keeping his shoulders parallel to the line of scrimmage. The lateral position of the backer to the ball carrier is critical to prevent a cutback and so this too is drilled on a regular basis.][B]

[The physical act of moving laterally keeping the shoulders parallel to the line of scrimmage is practiced nearly daily in agility drills. Blocking bags are often put in the players' paths so they can learn to keep moving in traffic without taking their focus off the developing play.][C]

If the play is a pass, the linebacker has to respond to another set of keys. Man to man coverage is cognitively simpler, but physically more demanding. [In man to man coverage the backer has to be taught and subsequently drilled on footwork to respond to a pass receiver's moves. The footwork is taught to the players individually, at first just having them respond to hand signals. The next step is to work one on one with a receiver, starting at a trot and progressively increasing the speed. Finally this will be done with a full complement of receivers and pass defenders. The footwork is practiced regularly in the agilities part of the practice throughout the season.][D]

In zone pass coverage the linebacker has to be coached to respond to keys provided by the quarterback and the pass receivers. Once the play has been determined to be a pass, the quarterback provides the initial keys. [While the quarterback is dropping back the linebacker has to be getting as deep as he can in his assigned zone. This is accomplished using crossover steps that are practiced daily in agility drills. When the quarterback stops and raises the ball to shoulder height in preparation to throw, the linebacker has to square up so that he can move in whatever direction the ball is thrown.][E] This skill is introduced in a one-on-one drill, initially without the ball being thrown, incrementally building up to the ball being thrown with a full complement of receivers and defenders.

The exact spot to which the linebacker drops is determined by the offensive formation and a read on the initial movement of the pass receivers off the line of scrimmage. (D. Mantyka, personal communication. March 15, 2004)

Part II.

_____ 1. Learn the factoring rule: $x^2 - y^2 = (x+y)(x-y)$.
Practice it until you can use it automatically. Then
practice it in questions like:

$$a^2 + 2ab + b^2 - y^2 = (a+b)^2 - y^2 = (a+b+y)(a+b-y)$$

and

$$x^4 - y^4 = (x^2 - y^2)(x^2 + y^2) = (x+y)(x-y)(x^2 + y^2).$$

_____ 2. When adding two fractions, if the denominators are the same,
add the numerators. If the denominators are unlike, find the
lowest common denominator.

_____ 3. Students learn that $n^{-1} = 1/n$ for any number n, and they
practice using that rule regularly. They also practice
questions like:

$$-6^{-1} = -1/6 \text{ and } (x+y)^{-1} = 1/(x+y)$$

so they do not get distracted by negative signs and additive
terms and use the rule incorrectly.

_____ 4. "I use simple arithmetic and algebra to rewrite $2x + 7x$ as $9x$.
Intelligence comes into play when the $2x$ and $7x$ are separated
by many other terms or appear on opposite sides of an equals
(=) sign. Then, one needs to be able to notice which terms
can be combined before they are combined. But even
the algebraic manipulations of expressions can become a
relatively easy exercise, given a lot of practice." (T. Kocurko,
personal communication, July 30, 2002)

_____ 5. If there are many terms to be combined with different
operations in an algebraic expression, put brackets around the
terms to be multiplied or divided to make sure you do them
before addition and subtraction.

38

Part III.

The final desirable outcome for the linebacker in football is to be able to play his position well in a game. This entails practicing basic footwork and doing agility drills in every practice. It also entails practicing these same drills man to man and then with various complements of offensive and defensive players in every practice.

The final desirable outcome for the mathematics practitioner is to be able to problem solve using mathematics. This entails regular practice of basic skills. It also entails regular practice of these skills in increasingly complex practice problems. What it does NOT entail is a lot of problem-solving without the regular repetitious practice of essential component skills. Solving large-scale interesting problems may be more appealing, but it does not provide adequate training for the real mathematics practitioner. That level of spectacularity must come later.

"Practice does not make perfect. Perfect practice makes perfect.

Once you've established a game plan based on your values, your goal each and every week as you prepare for the next game or event needs to be cutting down on practice errors. People in organizations should develop a fascination with what doesn't work. If you spend some valuable time concentrating on eliminating practice errors, you'll also eliminate a good amount of the second-guessing that goes on come performance time. Affirming and redirection are where organizations secretly outstrip the competition. Every mistake should be noticed and corrected on the spot. There's no such thing as a small error or flaw that can be easily overlooked."

- Taken from *The Little Book of Coaching* (Blanchard and Shula, 2001, pp. 46-47)

One of the most frequent comments from students when going over errors made on mathematics tests and assignments is, *oh that was just a stupid mistake.* We do not allow students to classify mistakes in this manner. We insist that the mistake be acknowledged as something important that must not be repeated. We seek to classify this error in a manner that will facilitate its correction. So instead of calling it *a stupid mistake,* we offer more useful categories like *multiplication fact error, long division process error, slip in concentration,* or *lack of automaticity.* Then we establish a practice procedure to avoid more instances of that particular error.

One of our means for building automaticity is a piece of software we developed called *Mathdrill* (May, Rabinowitz, Hart, and Larson, 1995). This program was designed to drill students in algebraic principles in order to help students respond accurately, quickly and ultimately, automatically. The program was based on the resource literature reviewed and mentioned in other sections of the book as well as on the memory findings summarized by Salisbury (1990) that are relevant to drill programs. These include using: small subsets of items to teach new concepts in order to reduce interference; spaced, rather than massed, practice to improve retention; and occasional review or reinstatements of earlier learned material in which the concepts are mixed to facilitate discrimination, retention, and appropriate use.

During the two-year interval between 1995 and 1997, we conducted a controlled experiment to test the effectiveness of Mathdrill in skills development. The results supported our hypothesis. (May, Rabinowitz and Mantyka, 2002, pp. 27-33)

Many math educators are content with a minimum level of competency in topic areas. Our experience has been that unless core algebraic skills are over-learned to the point of automaticity, errors in these skills occur in more complex problem-solving situations. Our experience supports the Blanchard and Shula principle cited earlier, "Practice does not make perfect. Perfect practice makes perfect" (2001, p. 46) and we require all our students to engage in perfect practice. If you want to teach or learn mathematics well, you must be prepared to do the same.

News item: more math specialists needed in Saskatchewan schools

"More Math Specialists needed in Saskatchewan Schools"

- Taken from *The StarPhoenix* (JOHNSRUDE, 2002, p. A15)

In the spring of 2002 there were several newspaper articles printed in Saskatoon's *The StarPhoenix* in response to the article, "Poor math skills stump experts" (Lang, p. A3). This article reported that "new statistics show Saskatchewan students have one of the lowest scores in the country."

For me this was like *deja vu*. For fourteen years prior to the appearance of this series of articles, I had been working on this problem in another province of Canada, Newfoundland. In the early 1990s I had delivered a workshop to mathematics teachers in Saskatoon at the invitation of the mathematics consultant for the Saskatoon public school board. At that time, given what the mathematics consultant told me was happening in curriculum development there, I had predicted what I saw reported ten years later.

The series of articles that appeared in *The StarPhoenix* laid the blame everywhere – on the lack of standardized testing, on the lack of communication between high schools and the university, on insufficient numbers of mathematics specialists in schools (see the cartoon on the left-hand page), on an out-dated curriculum, on poor reading skills of students. Conjecture after conjecture based not on data analysis, but on individual opinion. Yet educational policy is often set in this manner (*A Tale of Two Approaches to Policy Development: Evidence-Based Versus Knee-Jerk Anecdotal*, 2002).

On the Mathematics Learning Centre website, there is an article that I wrote for *The Telegram*, in which I outline an entirely different approach to this problem (May, 2002). In this article, I conclude that there is no quick fix, but that there are many, many things which individuals can do immediately at no cost and with no change in infrastructure, to improve the situation significantly. I encourage you to read this article in its entirety and judge for yourself where we should look for real solutions to the math problem.

"Both men and women have emotional needs. We all want to belong to a group. We all want to feel some worth, to know that people care about us and love us. When you can meet those needs for people, even in an aggressive, competitive arena, they'll respond with incredible effort."

- Frosty Westering, 73, head football coach, Pacific Lutheran University, U.S.A., taken from Where Pride Still Matters, *Men's Health* (Brooks, 2002, pp. 80-83)

The beginning of the academic year at the Mathematics Learning Centre is always tough. That is because almost all the students in attendance there at that time have just found out they failed the mathematics placement test. Most of these students have good grades in mathematics from high school, but we tell them that their skill levels in mathematics are at a grade six level or below. These students are angry and frustrated and justifiably so. These students have entrusted thirteen years of their education to their teachers and school curriculum developers only to find out that they have been duped into believing they would be well prepared for post-secondary mathematics. Or is it the university professors that are wrong? Are the standards being imposed by post-secondary professors of mathematics unreasonably high? Who are these students to trust now?

The staff at the Mathematics Learning Centre is fully aware of the emotional trauma that students are dealing with when they first attend class at our facility. For many of them, it is their first failure. Their need to upgrade their mathematics skills also sets them back a full year in their desired degree. This has significant time and financial implications for them. Furthermore, they are being told by us that they must go back and fill in the gaps in their knowledge of arithmetic processes including subtraction of whole numbers with borrowing, long division of whole numbers, all arithmetic operations with fractions, and so on. This just widens an already large credibility gap.

We understand that unless the students are given an opportunity to learn to trust us at the Mathematics Learning Centre, nothing useful will come of their time with us. But this process must begin with us. From day one, we make it clear to our students that we respect them as individuals. We understand that their requirement to study with us at the Mathematics Learning Centre is not easy for them to accept. We acknowledge that while they have under-achieved in mathematics, they have been misled in being made to believe otherwise by the educational system. We make a commitment to our students to work with them as long as they act on a similar commitment to work with us, no matter how much effort it takes. And we respect their choices whether or not we agree with them, but only after we have had an honest exchange of perspectives on the consequences of those choices.

In all of this, we are merely acknowledging the affective side of learning, just as Frosty Westering acknowledges the athlete's emotional needs in his quotation on an earlier page. In the study of mathematics as in athletics, the response of the individual is the same. As Frosty Westering puts it, "they'll respond with incredible effort." At the Mathematics Learning Centre it is essential that the students respond with incredible effort because that is the only way they will succeed given their level of mathematical knowledge at that stage in their education. At the same time, the individual's willingness to do so under those conditions is often a life-changing experience. From the myriad of testimonials we have on file from 18 years at the Centre, I include only a few below.

> The instructors for this semester were terrific. They were always there to lend a hand and let me know that they were there to help. Both the instructor and teaching assistant were consistent each class in helping when I needed it. The most impressive qualities were: (1) Their ability to explain each problem effectively; (2) They didn't think any problem I encountered was not worth tackling and in the process I didn't feel intellectually deficient. Thank you for a great experience! (*Course Evaluation Questionnaire*, Spring 2002. Math 102F, 103F, 104F, "Other Comments". St. John's, NL: Memorial University)

> Just wanted to write you and not only thank you for giving me the opportunities you did at MUN to get the math done, but also for understanding and caring enough to help . . . . if I learned one thing from that place is that I needed to work hard at math, and always need to review the basics and keep on top of things. Well so far, my efforts have paid large dividends. On my first test I received a mark of 85%, and only hope my perseverance will continue to bring me happiness. Anyway, I guess I just wanted to thank you for your help, and to let you know where I am. (A. McGrath, personal communication, Oct. 17, 2000)

> Thank you for your assistance and support . . . . I found your encouragement and trust tremendous and those

times where I wasn't prepared to pre-test, but you helped
me anyway – those times pushed me to work harder yet
because I didn't want to let you down. I know some people
and students in particular question the validity of the
Mathematics Learning Centre. I'm not one of them. Math
102-103 has helped me overcome an immobilizing fear of
math, and if you can overcome your greatest fear then little
else can stop you. Thanks for all that you're doing here and
God Bless. (L. Shoemaker, personal communication, Dec.
2, 2003)

All these individuals overcame their initial aversion to being at the
Mathematics Learning Centre by identifying with a common purpose.
This was a result of our responding to the individual's need "to feel some
worth, to know that people care about us," as Frosty Wethering puts it, and
the returns have been staggering, both for the students and for the staff.

"All men need to understand that dreaming alone isn't enough. Dreaming can be a way of kidding yourself. I spend a lot of time getting my players' heads out of the clouds and getting them to deal with reality. What you need to do is to make choices and work in ways that are consistent with making your dreams come true. A lot of my players face temptations of partying and late hours. That's why I hold practice at 5:30 a.m. I try to structure in methods and a program to help them with self-discipline, which ultimately helps them become successful."

- John Chaney, 69, basketball coach, Temple University, U.S.A., taken from Where Pride Still Matters, *Men's Health* (Brooks, 2002, pp. 80-83)

Realistic goal setting is an important element of success. So is self-awareness. It is simply not true that everyone is equally talented or has the same financial and other support resources at hand to achieve his/her goals. There is nothing wrong with shooting for the stars as long as you are willing to accept a lot of failures along the way and have the determination and self-confidence to persevere in spite of the failures. Most individuals who are regarded as high achievers have suffered nine failures for every one success. Not every one has the capacity to deal with that reality. As John Chaney put it in the quotation on the left-hand page, "Dreaming can be a way of kidding yourself" (Brooks, 2002, pp. 80-83).

Another key element of success in achieving high standards of performance is self-discipline. Climbing the staircase to the stars can be repetitive and boring – the next stair looks much the same as the last one. In athletics, in music, in mathematics, it's all about basic skills, repetition, practice. And it takes an incredible amount of self-discipline to keep at it and at it and at it. One way of assisting yourself in being disciplined is to develop structures in your routine that get around your personal weaknesses. So, for example, John Chaney schedules practice at 5:30 a.m. because he knows that if practice starts later, his players will be more tempted to keep late hours. There is also an unexpected spill-over benefit from forced distasteful routines like early morning practices. It is that many individuals discover that they actually enjoy the rest of the day more than they would have otherwise because by the time the rest of the world is barely getting started, you have already completed a half-day's work. This can be an incredibly energizing factor in the rest of your day.

At the Mathematics Learning Centre, we do not hold practices at 5:30 a.m., but we do hold classes and test centres between 8 a.m. and 10 p.m. every weekday except Friday. These sessions are compulsory for our students. We have also developed very structured planning templates to assist our students in both realistic goal-setting and daily practice sessions in mathematics. Initially, we meet individually with our students to assist them in mapping out their program in mathematics. We solicit information from our students about other commitments in their lives.

These must be factored into their plan for study of mathematics in order to be realistic. This includes everything from employment commitments, care-giving obligations, medical conditions, transportation issues, financial constraints, and so on. Throughout the academic semester, we meet weekly with our students individually to assess their program progress to date, to discuss behavior that may have contributed to non-achievement of satisfactory progress, and to adjust their plans and/or goals accordingly if necessary.

We have observed that success in our upgrading program in mathematics generates success not only in further mathematics courses but also in English courses and overall university graduation rates. Because of this we have concluded that the attitudinal shifts that result following successful completion of our program are more important to the learner's success than the actual mathematical content of our program. Success comes with knowing how to structure the learning experience and that applies to everything, not just mathematics and not just academic pursuits.

WEEKLY PLANNER

Mathematics Learning Centre

Semester: __Spring 2006__   Name: _____

Week	Start Date	Units to Study	Module Tests			Passed
			1st Write	2nd	3rd	
1	May 8	R.S. 1-5, Unit 1				
2	May 15	Units 2, 3				
3	May 22	Supp. I, Module 1 Review, Unit 4	SRF 1			
4	May 29	Units 5, 9				
5	June 5	Units 6, 18				
6	June 12	Unit 19, Module 2 Review, R.S. 6-12	SRF 2			
7	June 19	Supp. II, Units 7, 8				
8	June 26	Supp. III, Module 3 Review, Unit 10	SRF 3			
9	July 3	Supp. IV(i), Units 12, 13				
10	July 10	Units 14, 15, Supp. IV(ii)				
11	July 17	Units 16, 17, Module 4 Review	SRF 4			
12	July 24					
13	July 31					

TEST CENTRE TIMES   **CLASS PERIOD**

Tuesday   12:00 p.m. – 4:00 p.m.   **Slot 3A**

Thursday   12:00 p.m. – 8:00 p.m.   M, W, R, F - 10:00-10:50 p.m.

NOTE:
1. Do not come to class without **ALL** your books. Bring with you to **EVERY** class your textbook, student manual, course manual, and **ALL** the work for the module you are currently working on and for which you require a signature.
2. The class periods are times for you to work, to get help and to obtain any necessary signatures.
3. You must have your work checked, and obtain the required signature, before you write a module test. This must be done during a class period. Signatures will not be given during Test Centres.
4. Tests written on Tuesday are returned on Friday. Tests written on Thursday are returned on the following Monday. Failed tests **MUST** be collected during class times.
5. 1st attempts of module tests must be written, at the latest, by the Thursday Test Centre of the weeks indicated above. If you do not write the 1st attempt by the week indicated above, you will forfeit your 1st attempt and therefore the highest score possible on the test would be the minimum required to pass. Students should aim to write their 1st attempts the week before the above deadlines.
6. **No graphing or programmable calculators are allowed in the Test Centre**. Therefore, students should use only scientific calculators at all times.
7. Last Test Centre: August 3, 2006.
8. Last Class: August 3, 2006.

"U2 and Denise who?

Halifax native and cello sensation Denise Djokic and her Stradivarius open for the rock band Train tonight at the Grammys"

- The Globe and Mail
(Mandel, 2002)

This article embodies the unexpected – a young, successful, prodigious, classical cellist playing a segment of a Bach cello suite as a lead-in to a soft-rock band's *Drops of Jupiter*, nominated for Best Rock Song at the annual pop music Grammy Award ceremony.

In the article, Djokic is quoted as follows regarding her childhood friendships: "My devotion to practicing daily was kind of like their devotion to going to soccer practice for two hours. It didn't really set me apart, although I knew at that point how serious I was about music" (Mandel, 2002).

Many great musicians spend hours in repetitious practice of scales; great soccer players spend hours in repetitious practice of heading. Each is committed to excellence and understands that attaining excellence is not primarily about spectacular activities. The end result is indeed spectacular but the preparation can at times seem endlessly tedious.

So, too, it is with attaining useful skills in mathematics. Commitment to the outcome, the self-discipline to do what is required, and an appreciation for the possibility of the unexpected – these are all key elements to sustaining the necessary effort.

"Holy roller role for Jets rocker

From fronting a band to treading the boards as Judas is something of a stretch. But as Dave Gleeson tells Chad Watson, he has thrown himself into the role – body and soul"

- Taken from *The Newcastle Herald* (Watson, 2002)

When Dave Gleeson accepted the role of Judas in a Newcastle Dramatic Art Club production of *Jesus Christ Superstar*, his fans knew him as an irreverent hard-rocker from The Screaming Jets. But Dave Gleeson also had a history in a strong Catholic family as a former choirboy from St. Joseph's Primary School, and The Screaming Jets had played their farewell gig 18 months previously. Dave Gleeson was looking for work.

Unlike the economy of 50 years ago, people can no longer expect to have the same job throughout their adult life. We have already begun to experience this shift in lifetime employment patterns as traditionally resource-based regional economies shift to more technologically-based operations. This has certainly been the case in Newfoundland with the collapse of the fishery and with the decline of the pulp and paper industry in our province. This has forced many adults to consider re-entering the educational stream to train for employment in technology-based industries. Mathematics is a key subject in this domain and for many of these adults, it was the subject that played a significant role in driving them out of the educational system in their youth.

In *The Little Book of Coaching*, Blanchard and Shula cite the following principle as being pivotal to success: "Be prepared with a plan and then expect the unexpected and be ready to change your plan" (2001, p. 60).

In athletics, we have already seen a fine example of this in the response of the New England Patriots in their 2001-02 football season when they lost their $100 million quarterback to an injury early in the season (see "Champions! Patriots win Super Bowl, 20-17, on Vinatieri's last-second kick," pp. 30-31). Had the coaching staff of the Patriots not been prepared with a plan and expected the unexpected and been ready to change their plan, they never would have managed to win the Super Bowl that season.

So it is with Dave Gleeson. With his decision to accept the role of Judas in *Jesus Christ Superstar,* Dave Gleeson acknowledges this type of resourceful adaptation to situations as a key to his success in the music industry. His decision-making is not bound by image or immediate past working environments. Rather, he accepts the unexpected opportunity and works hard in rehearsals to make it work well. Of this, director Carolyn Lind says:

> When Dave first came in he was blown away because it
> was so different to what he's used to. But now he's going
> [doing] remarkably well . . . he's getting along with the
> cast and fits in perfectly. You can see Judas really coming
> out. It took a while before he came to grips with playing
> someone else other than himself. He also had to realize
> that, at times, there are 37 other people up there with him,
> but he's more than met the challenge. (Watson, 2002, p. 31)

At the Mathematics Learning Centre, we also acknowledge that adaptability is a key to success because of our work with adult learners. Of the experience, our students say:

> I would like to say thank you for the positive feedback that
> accompanies each of my assignments and tests when they
> are returned to me. Math has always been very difficult to
> me and the positive reinforcement is greatly appreciated.
> (S. Howell, personal communication, Mar. 1, 2002)

The best aspect of this course was the quality of instruction and encouragement. The course manual as well is fantastic. For the first time in my life I actually enjoyed my math course. I accomplished what I set out to do with excellent guidance from both the manuals and the instruction. (*Course Evaluation Questionnaire*, Spring 2002. Math 102F, 103F, 104F, "Question 21." St. John's, NL: Memorial University)

For these amathematical students who found themselves contending with a math skills upgrading program, this is a major turnaround for them. They have had a life plan and they have been forced to revise that plan. They have faced that reality head-on and accepted the challenge of the consequences. They have chosen to trust our judgment and worked very hard at the mathematics we have assigned them. They have succeeded at what they set out to do and are now in a very strong position to get on with their lives.

If you want to succeed, you must be willing to do no less. Perhaps you had a plan that did not include too much mathematics. Or perhaps you had a plan that did include mathematics but you did not expect it to be so hard. Now you find that you have to revise that plan, either to include mathematics or to work harder at it than you expected. No matter what field of endeavor you look to for inspiration, you will find countless examples of success stories that embody the willingness to do what you must now do. I challenge you to go look for these instances no matter who your hero(ine) is. That is, after all, what the technology of the internet is meant to do for us, isn't it? It is meant to expand our horizons beyond what is right before us.

"A few gems amid the trashy Grammy divas

Canadian winner Nelly Furtado makes a case for not letting it all hang out in a gown fit for a pop princess"

- *The Globe and Mail*
(Pearce, 2002, p. 12)

Image is important in today's world. It affects the way we feel about ourselves and it affects the way others perceive us. We can choose an image for ourselves that will make us blend in or one that will make us stand out. We can choose an image that will lead our peers to associate us with a certain group of people – athletes, the social crowd, the serious students.

Of course you do not have to look the part to play the game. Our image is only a small part of what is really happening in our lives, and our image does not have to match our behavior.

Historically, academically inclined individuals have not had a widely appealing image. Einstein was no Brad Pitt, although in the movie "A Beautiful Mind," Russell Crowe made a fine piece of eye candy. Casting Russell Crowe as the lead in this film was undoubtedly designed to broaden the film's appeal beyond the academically inclined, and it worked. This movie did very well at theatre box offices.

Lots of young people today face the dilemma of feeling an inclination to do mathematics but a disinclination to be identified with the image their academically inclined peers present. This is a perfectly legitimate concern.

In the entertainment industry, Nelly Furtado constantly faces decisions about her image. In her profession, image has been paramount for many performers, for example, The Spice Girls, The Backstreet Boys, Britney Spears, Marilyn Manson. Yes, there was talent, but there was a lot more image. For the Grammy Awards in 2002, Nelly Furtado made a bold decision to dress more conservatively than her competition and it paid off. Of her, reporter Tralee Pearce writes in the article cited on the left: "Amid the tired hoochie-girl looks, Furtado, Grammy winner for best Female Pop Vocal Performance, stood out in her embroidered, gauzy gown and matching wristband – the best of the night. Sure, she's got great abs under it all, but her now-characteristic restraint is refreshing" (2002, p. 12).

Students who like to do mathematics are similarly challenged every day. It's a matter of self-confidence to be able to do what you want to do and dress the way you want to dress. The two do not have to go hand in hand by standards that your peers set – it's your choice.

"Noted for her elegance and sense of style, Eileen could be counted on to cause some judicial heads to turn. Always tasteful, she had the tenacity and self-confidence to dress with flair. Her courtroom ensembles ranged from chic suits to funky skirts. Her unmistakable panache changed the public image of the judiciary."

- Taken from "Lives Lived: Eileen Nash, Mother, friend, judge of Alberta's Court of Queen's Bench . . ." *The Globe and Mail* (Grieve, 2002, p. A18)

The written eulogy to Eileen Nash cited on the left-hand page was composed by her friend, Kathy Grieve, former chair of the Edmonton Police Commission. In this eulogy Kathy Grieve acknowledges Eileen's determination to be brilliant and breathtaking, to break barriers. This is a call to all of us as individuals, not just to those who will pursue careers in the judicial system, or as mathematical practitioners.

In a book entitled *The Little Book of Coaching*, written by renowned football coach Don Shula and industrial psychologist, Ken Blanchard, the following principle is cited: "Character is the sum total of what you believe and how you act" (Blanchard and Shula, 2001, p. 28). Shula makes the point that character is just as important as ability to being successful. This fact cannot be ignored in the teaching and learning of mathematics. If classroom experiences are limited to mathematical content, no real growth will occur. This is a tall order for both students and teachers. It means that the successful teaching/learning relationship must be completely engaging – engaging the mind, the heart, the spirit. Nothing less will do.

" 'The best part about being a member of Great Big Sea is the opportunity to perform traditional music around the world,' says Hallett; however, he's quick to add that 'the worst part is the tedium of getting there.' "

- Taken from "The members of Newfoundland's most popular foursome are this year's alumni of the year" (Etchegary, 1999)

From the article by Victoria Etchegary, we quote further:

> Great Big Sea's list of awards and accolades since their
> arrival on the music scene in 1991 is impressive. Winning
> the East Coast Music Awards (ECMA) Entertainer of the
> Year Award four consecutive years (1996-99) has proved
> beyond a doubt that Newfoundland traditional music is
> a viable and important part of the music industry. The
> group has also won six other ECMAs, including Album of
> the Year in 1998 for *Play*, and six awards from the Music
> Industry Association of Newfoundland and Labrador.
>
> The distinction of Alumni of the Year is a new award for
> this group, one they are pleased and honored to accept.
> "Winning Alumni of the Year is a much appreciated nod
> from a learned group that, as my transcript marks will
> show, do not hand out praise lightly," says Doyle. "I am
> sure we are in good company as there are many deserving
> of this recognition. I am honored."
>
> All four group members met, and assembled as Great Big
> Sea while they were arts students at Memorial (University).
> (Etchegary, 1999)

For many people, the study of mathematics can definitely be described
as Hallett does on the left-hand page, "the worst part is the tedium of
getting there" (Etchegary, 1999). Our experience working with under-
achievers on a one-to-one basis has led us to believe that the largest
factor contributing to lack of success in school mathematics is attitudinal.
Students do not expect to have to work at mathematics and those who find
it difficult are either unmotivated or paralyzed by mathematics anxiety. At

the Mathematics Learning Centre attitudes are changed because we: (a) are prepared to treat each learner as an individual with interest and respect; (b) provide a supportive learning environment tailored to individual needs; and (c) require steady, regular completion of work at a pace we set. This not only facilitates the successful completion of our course, but also the successful completion of regular post-secondary pre-calculus and calculus courses that follow. For example, 73 percent of the students who completed our program passed the pre-calculus course on the first attempt. This is in stark contrast to the 20 percent pass rate of the comparable 1988 sample scoring below 50 on the placement test who were sent directly into the pre-calculus course. Furthermore, 67 percent of our students who completed the pre-calculus course enrolled in calculus. Of these, 74 percent passed on their first attempt and all passed by their third attempt.

Passing courses in mathematics is not as spectacular as performing in concerts around the world, but it is definitely a means to better employment opportunities. No matter what the endeavor, there is no escaping the tedium. As Doyle says, "it's sometimes because of the cumbersomeness of traveling that the band can cherish the 90 minutes on stage" (Etchegary, 1999). Having attended their concert in St. John's recently, I can certainly attest to that.

"Do something every day
that you don't want to do;
this is the golden rule for
acquiring the habit of doing
your duty without pain."

*****

"One must learn by doing
the thing; for though you
think you know it, you have
no certainty, until you try."

- (Teachers, May 27/28 & April 20, 2006)

"Changes in emphases require more than simple adjustments in the amount of time to be devoted to individual topics; they also will mean changes in emphases within topics. For example, although students should spend less time simplifying radicals and manipulating rational exponents, they should devote more time to exploring examples of exponential growth and decay that can be modeled using algebra."

- Taken from the *National Council for Teachers of Mathematics Curriculum and Evaluation Standards for School Mathematics* (1989, p. 151)

The *Standards* document referred to in the quotation on the left-hand page has been the primary reference in all curricula revisions made in North America since its publication in 1989.

The flavor of the quotation on the left-hand page is characteristic of the whole document. Throughout the K-12 curricula, this document recommends an increased emphasis in problem solving and a decreased emphasis in numeric and symbolic manipulation. While this is not necessarily a bad thing, taking it in the extreme, as has been the case in the implementation of the new curricula, is disastrous.

The reason for the disastrous effect upon the learning of mathematics comes from cognitive psychology. The following excerpt from "Teaching the Rules of Exponents: A Resource-Based Approach" by May, Rabinowitz and Mantyka, accounts for this outcome.

> Norman and Shallice (1986) provided a fairly simple framework to deal with mathematical problem solving and the relationship between reasoning and remembering. They suggested that the supervisory attentional mechanism is the limited resource and assume that deliberate attentional resources are employed in tasks that involve planning or decision making, troubleshooting, ill-learned or novel sequences, dangerous or technically difficult components, and the inhibition of strong habitual responses . . . . Perhaps all the attention-demanding characteristics are involved in solving difficult mathematical problems.
>
> Norman and Shallice also suggested that the automatization of skills plays an important role in problem-solving tasks which require attentional resources. Within the context of resource theories, *skill* seems to represent any mental or physical response that is performed in completing a task. It is assumed that fewer resources are required as skills become more rapidly executed and that over-learned

skills become *automatic* and no longer require resources. Automatic skill use is thought to be implicit, rather than explicit, and, therefore, less likely to be forgotten (see Berry & Dienes, 1993). One way this could happen is that over-learned skills could be directly retrieved from memory rather than constructed (Logan, 1988). For example, with sufficient practice most children know that $2 + 3 = 5$ without needing to count and many high school students recognize that $(x-a)(x+a) = x^2 - a^2$ without multiplying the terms in the expression. At a more advanced level, a skilled high school student would automatically add the exponents, rather than first retrieving and then applying the addition rule when presented with $x^b x^c$. Automaticity is probably the most appealing resource-based concept from the perspective of a mathematics educator because over time, a child and later, a young adult, gets to practice a large number of skills repeatedly. Some of these highly practiced skills (e.g., adding, subtracting, multiplying, dividing, simplifying fractions, factoring) are useful in a large number of mathematical tasks and should be associated with developing expertise. The absence or slow execution of many of these basic skills in our remedial mathematics students is certainly an indicant that something went wrong. (May, Rabinowitz and Mantyka, 2001, pp. 68-69)

These paragraphs highlight two major flaws in teaching mathematics the way the *Curriculum and Evaluation Standards for School Mathematics* advocates.

Firstly, in curricula guided by the *Standards* document, new skills are taught in the context of problem solving. This directly contravenes the principles of cognitive psychology outlined in paragraph one of the quotation (on p. 67). Because new skills represent either novel or ill-learned sequences that also may be technically difficult, they place unusual demands on capacity. Thus, if a new skill is being taught, it is best to minimize other demands until that skill is mastered. For example, word problems should not be introduced until the relevant mathematical skill(s) is/are mastered.

Secondly, in de-emphasizing numeric and symbolic manipulation, skills are not being practiced enough to achieve automaticity. Demands on attentional resources are less if the level of relevant skills is higher. Therefore, skill enhancement should reduce demands on capacity and foster complex problem solving. Three ways in which relevant skills can be improved are: practice, the substitution of efficient skills for cumbersome skills, and the use of shortcuts to reduce steps. None of these are emphasized in modern curricula guided by the *Curriculum and Evaluation Standards for School Mathematics*.

For teachers and students alike, this is easily corrected. Teachers must require more practice of basic skills outside the context of problem solving, and students must be willing to do the practice.

"Basic skills remain key to the game

SOCCER

Sorry, but we need to get technical.
In the thrill of the chase and the power
of the shot, it is sometimes forgotten that
soccer is a game requiring fundamental
skills such as control, passing, tackling,
heading. Quick feet are like quick hands
in tennis – crucial but you still need a strong
forehand, backhand and serve if you
want to become an outstanding player."

- Taken from *The Globe and Mail*
(Palmer, 2002, p. 56)

We are back to drawing the analogy between basic skills in athletics and basic skills in mathematics. In this instance, we are going to analyze how the current emphasis on technology in modern curricula affects skill development. In particular, we will focus on calculator usage.

The following excerpts are taken from Grade 3 curriculum documents recently produced for the Department of Education for Newfoundland and Labrador. This curriculum is based upon the *Curriculum and Evaluation Standards for School Mathematics* (National Council for Teachers of Mathematics. 1989) and The Foundation for the Atlantic Canada Mathematics Curriculum.

Principles:

2. Children learn mathematics through worthwhile activities as they explore, communicate, and reflect on important mathematical ideas and procedures in a variety of contexts.

. . . .

5. Children learn mathematical skills through problem solving, and practice these skills through independent work.

. . . .

9. The tools of technology – computers and calculators – and concrete objects help children learn to solve problems in ways that will prepare them for the future. (Wortzman *et al.*, 1996, p. x)

General Curriculum Outcome B: Students will demonstrate operation sense and apply operation principles and procedures in both numeric and algebraic situations.

Suggestions for Assessment:

B11.1   Ask the students to add mentally as you draw numbers from a bag, and to stop you when the sum has passed 40.

B11.2   Have the student make a list of calculations involving 2- and 3-digit numbers that would be quicker to do mentally than on paper or with a calculator.

B11.3   Ask the student to describe a strategy for calculating $48 - 9$ (or $76 + 11$) mentally.

B13.1   Ask the students to show how he/she would use a calculator to find $4234 + 187$. (*Mathematics: Primary*, 2000, pp. 173, 213, 215)

I have not deliberately avoided the selection of suggestions for assessment that actually entail producing an answer to an arithmetic sum involving 2- and 3-digit numbers. There were no such suggestions for assessment. In this excerpt from the curriculum guide, we see the extreme implementation of the changed emphases outlined in the *Standards* that is characteristic of modern curricula. Firstly, we note several examples in the suggestions for assessment that apply the principles of exploring, communicating, and reflecting, as opposed to doing. Secondly, we see an explicit declaration of learning mathematical skills through problem solving that was refuted in the cognitive psychology literature in the previous section of this book. Thirdly, the suggestion for assessment that entailed extending the procedure only by one place holding digit (e.g., the 3- and 4- digit sum listed at the end of the quotation), was a suggestion for assessment using a calculator. I do not believe that adding a 3-digit number to a 4-digit number, even with carrying, so overloads a child's attentional resources that calculator usage is called for. This merely fosters in the child a lazy reliance on technology to circumvent mental arithmetic. If this is the best use of technology that can be made in this

Grade 3 curriculum, then it is clearly not needed, and can only lead to a head start in the development of bad habits.

The young soccer player who believes s/he can develop the skills necessary to play soccer by playing *FIFA 2004* by Electronic Arts, is not likely to be a welcome member of any soccer team.   So it is with mathematics. A familiarity with mathematics is nice. Understanding mathematical concepts is fine. But to participate in the work force as a mathematical practitioner at any level, or to proceed to post-secondary education involving the use of mathematics, there is an expectation that the individual can do mathematics. This does not come with mere exploration and reflection. This comes with hours of practice, much of which is repetitive.   Adequate practice to achieve automaticity is not the result of independent study and certainly does not involve the use of a calculator for simple whole number computations at a Grade 3 level.

" 'So far so good,' left-handed reliever Graeme Lloyd said. 'I don't see anything that hasn't been done right, but that doesn't really surprise me,' the nine-year veteran said. 'After all, spring training's about getting your fundamentals done, about working on our plays and getting ready.' "

- Taken from "BASEBALL MONTREAL EXPOS **Players show no sign of being lame ducks,**"
The Globe and Mail (Blair, 2002, p. S3)

At the time of this newspaper report, the existence of the Montreal Expos baseball team after the 2002 season was highly uncertain. At spring training camp, however, the focus was on fundamentals, and not on razzle-dazzle game strategy.

The importance of fundamentals in any major endeavor cannot be over-emphasized. It was the underlying strength of the New England Patriots in their journey to their first Super Bowl triumph in 2001-2002. It was the basis for the 13-0 season of the Titans in the memorable movie *Remember the Titans* (Bruckheimer, Oman, and Yakin, 1989). It is the basis of the success of the students at the Mathematics Learning Centre in doubling their university graduation rate. It was the focus of the Montreal Expos coaching staff during this pivotal spring training camp.

It is amazing that curriculum developers seem to be the only group that does not acknowledge this fact. After all, the *Curriculum and Evaluation Standards for School Mathematics* (National Council of Teachers of Mathematics, 1989, p. 150) postulates the standard for senior high school algebra to be representation, interpretation, appreciation, and operations, in that order of importance. When I was writing this book in Australia, I gave a talk to a group of professional women (Newcastle Enterprising Women). During the talk I asked the women to write and submit to me something about their perception of whether or not there was a math problem in Australia. Of twenty-four women, only one woman said she had no basis on which to form an opinion and only one other woman believed there was no problem. (She was the only woman in the group to have a formal affiliation with the K-12 school system.) The other twenty-two women wrote extensively on their perception of the situation in Australia, and recorded such comments as:

> I am an accountant and we are highly reliant on computers.
> When I went to school (1969-1981) we did not use
> calculators until the final years of high school. We rote
> learned our times tables, algebra, calculus, statistics,
> trigonometry, etc. Learning how to do these things
> manually has helped develop my analytical skills. The

younger professionals have spent their years at school using calculators so their analytical skills do not appear to be anywhere near as developed. (Mantyka, 2004)

I have ten-year old twins, a girl and a boy. They are in separate classes. The boy had a very traditional teacher last year who believed in times tables, etc. every day at school. He excelled and gained great confidence in that year. My daughter is not math orientated, is nowhere near as confident as my son. Her teacher was not as persistent although capable. As a mother I encourage them very much as I do not have confidence with math. I really do not remember being helped much with math by my teachers especially at age 13 – my high school. For some reason I never felt confident at all. Now in my 40s I have my own business that really would benefit if I had confidence with math. You are an inspiration and made me review the importance of math in our lives. Many mothers I know just have so much trouble helping the children with math. It won't be long before I won't be able to help them. (Mantyka, 2004)

When our province was developing and piloting our new mathematics curriculum based on the *Curriculum and Evaluation Standards for School Mathematics*, many, many experienced teachers expressed grave concern about the lack of skills proficiency it would foster. Curriculum developers ignored this feedback. The University expressed its concern with the performance of the pilot students and the decline in participation in mathematics courses at university from the graduates of the new curriculum. Curriculum developers ignored this feedback. High school performance in the first graduating class was anticipated to be so poor that the provincial public exam was cancelled. There was a huge public uproar over this. Still curriculum developers ignored this feedback.

By 2007, all students in the Newfoundland K-12 education system will be victims of this highly flawed curriculum. There were massive data sets demonstrating its weaknesses. There was research in cognitive psychology outlining the fundamental flaw in the principles underlying the new curriculum. The majority of professionals in the field believed the new curriculum would fail. Students and parents knew the new curriculum did not make sense. Still we now have the devastating outcomes documented on my website (Li, 2003; May, 2003). Participation rates have dropped as much as 18% and pass rates have plummeted in one group from 70% to 43% and in another from 89% to 60%. It is hardly heartening to be able to say to curriculum developers now that they were warned. The tragic victims of this folly have been our children.

The Montreal Expos, however, are still a team enfranchised in Washington as the D.C. Nationals.

"It is important to separate 'outcome goals' – the big stuff – into 'process goals' – the little stuff you can accomplish every day. Larger goals, like winning championship titles, can often seem abstract and distant, but when you break them into things you can do every day, your dreams gain power, and it gives you a sense of purpose and control. By doing it that way, I've also learned to enjoy not just reaching the goal, but the daily pursuit of it."

- Jay Martin, 53, soccer coach, Wesleyan University, U.S.A., taken from Where Pride Still Matters, *Men's Health* (Brooks, 2002, pp. 80-83)

There are two important issues here, and the first is not complex.

Tackling large projects can be paralyzing unless the large project is broken down into more immediately attainable objectives. When things are going well, it is easy to feel encouraged and to continue to work hard at something like mathematics. But when you are facing a brick wall, it is easier to retreat than it is to attempt to scale the heights of that wall, especially when you anticipate many such brick walls to come. Often, when a student at the Mathematics Learning Centre is having difficulty with a particular task in his/her program, that student loses focus because of the anxiety generated. In those instances, we frequently suggest that the student back away from the brick wall and regroup with the assistance of a program planning template similar to the one illustrated on p. 51. This enables the student to concentrate on one troublesome topic in mathematics at a time by assigning each topic to specific future days and weeks.

That's what Jay Martin is talking about on the left-hand page. He calls the immediately attainable objectives process goals and the large project an outcome goal, but it is the same principle being applied to athletics.

The second issue is more complex because it has nothing to do with planning and everything to do with cognitive psychology.

Contemporary curriculum developers often undervalue practice. They focus instead on "pursuing open-ended problems and extended problem-solving projects" (National Council of Teachers of Mathematics, 1989, p. 70). Yet, there have been arguments made in the literature of cognitive psychology that support the importance of practice, indeed, declare it to be of paramount importance in the effective pursuit of open-ended problems and extended problem-solving projects. We cite some of this literature here:

Campbell and Xue (2001) have demonstrated that differential amounts of practice over more than a decade facilitates arithmetic skills. These scholars presented 72 students registered in undergraduate and graduate programs 360 arithmetic questions. Participants were required to add three 1- or 2-digit numbers, divide 2- or 3-digit numbers by single-digit numbers, subtract 2-digit numbers from 2-digit numbers, and multiply 2-digit numbers by 1-digit numbers. The number of these problems participants successfully completed in 15 minutes was predicted by reported calculator use before entering university and mean reaction time to answer simple arithmetic (e.g., 4×3) questions. Presumably early practice in single-digit arithmetic led to faster access to basic facts which transferred to more complex arithmetic while access to calculators reduced the number of opportunities to practice more complex arithmetic.

In addition, Haverty (1999) has demonstrated that concentrated practice facilitates complex problem solving. Haverty trained seventh graders to generate either the 17 or 19 multiplication tables, but not both the 17 and 19 tables. Following their mastery of these facts, the children were presented inductive reasoning problems at a variety of difficulty levels by providing them with tables, each of which contained six x, y number pairs, and requiring them to describe the mathematical relationship between the variables represented by the numbers in each table. The prior learning of the relevant multiplication facts facilitated solving the most difficult problems [e.g., $y = 17 \ (x + 1)$]. Haverty's finding is consistent with the assertion that mathematical discovery is dependent on prior skill level, and consistent with our contention that skill development usually is a consequence of practice over extended time periods that results in both the availability of and speedy access to facts and procedures.

This research in cognitive psychology makes a mockery of the "move away from a tight focus on manipulative facility to include a greater emphasis on conceptual understanding, on algebra as a means of representation, and on algebraic methods as a problem-solving tool" (National Council of Teachers of Mathematics, 1989, p. 150). One can no more learn to do mathematics well by immediately immersing oneself in problem solving than one can learn to play soccer well by immediately concentrating on playing games. As Jay Martin puts it, it's important to separate outcome goals into process goals. In so doing, "your dreams gain power, and it gives you a sense of purpose and control" (Brooks, 2002, p. 78).

"Low marks sending N.S. pupils back to mental arithmetic

Nova Scotia schools are going back to the basics of arithmetic after Grade 5 students across the province failed a standard mathematics test – achieving an average mark of 42 per cent."

- Taken from *The Globe and Mail*
(Cox, 2002, p. A6)

The newspaper article quoted on the left-hand page goes on to say that the Education Minister does not blame teachers or the school system for the results.

> "But she said the elementary school test results show that changes have to be made to allot more time for math and language skills and to improve the way in which those subjects are taught."

The author of the article also states that

> "The elementary school math test was done to assess the effectiveness of a new curriculum introduced five years ago."

This new curriculum would have been the result of the same Foundation for the Atlantic Canada Mathematics Curriculum that generated the new curriculum for Newfoundland. This new curriculum for Nova Scotia would also have adhered to the principles of the *Curriculum and Evaluation Standards for School Mathematics* (National Council of Teachers of Mathematics, 1989).

The position of the NCTM *Standards* on "Number/Operations/ Computation" in the Grades 5 to 8 mathematics curriculum is outlined below:

> INCREASED ATTENTION . . . .
> - Developing number sense
> - Developing operation sense
> - Creating algorithms and procedures
> - Using estimation both in solving problems and in checking the reasonableness of results
> - Exploring relationships among representations of, and operations on, whole numbers, fractions, decimals, integers, and rational numbers
> - Developing an understanding of ratio, proportion, and percent

. . . .

DECREASED ATTENTION . . . .
- Memorizing rules and algorithms
- Practicing tedious paper-and-pencil computations
- Finding exact forms of answers
- Memorizing procedures, such as cross-multiplication, without understanding
- Practicing rounding numbers out of context

(National Council of Teachers of Mathematics, 1989, pp. 70-71)

It is easy to see that a curriculum that is guided by these standards is unlikely to produce students who will be able to compute without the aid of a calculator. Of course increased attention to the use of technology for computation and decreased attention to paper-and-pencil computation is another principle adhered to in the *Standards*.

When I was writing this book in Australia, I asked twenty-four professional women whom I was addressing whether or not they perceived there to be a math problem in Australia, and if so, how they would characterize that problem. Twenty-two of the twenty-four answered yes to my query, and fourteen of the twenty-two said the problem was an over-reliance on calculators. I had not mentioned calculator usage in my talk at all. (Mantyka, 2004)

Clearly calculators play an important role today in facilitating speedy calculation of unwieldy numbers. The problem is that youth today believe that 42.5×100 is an unwieldy calculation, whereas anyone with a basic understanding of place value can immediately produce an answer to that question. In the primary and elementary classroom where calculator usage is rarely required for curriculum objectives, it is difficult to achieve the appropriate balance for its use. It is so much easier for both the student and the teacher to reach for the calculator instead of engaging in the painful thought processes and the tedious practice that comes with developing a quick mental facility with numbers. This is the danger, and we see in the newspaper article quoted on page 82, that Nova

Scotia students were suffering the consequences only five years after the introduction of the new curriculum.

At the Mathematics Learning Centre, we insist that students develop a facility with mental arithmetic long before they are allowed to use a calculator. In previous sections we have cited research in cognitive psychology which directly relates a lack of computational facility with calculator usage. Given the performance data and the scientific research in cognition, it seems pure folly to include calculator use in the primary and elementary grades. Yet we read over and over all through the *Standards*, statements like the following:

> The proposed algebra curriculum will move away from a tight focus on manipulative facility to include a greater emphasis on conceptual understanding, on algebra as a means of representation, and on algebraic methods as a problem-solving tool. For the core program, this represents a trade-off in instructional time as well as in emphasis. For college-intending students who can expect to use their algebraic skills more often, an appropriate level of proficiency remains a goal. Even for these students, however, available and projected technology forces a rethinking of the level of skill expectations. (National Council of Teachers of Mathematics, 1989, p. 150)

I wonder when we, as parents and math educators, will be released from the shackles of this misguided document that is currently dominating all curriculum development in North America, so that our children might be released from the shackles of failure cited in the newspaper article.

"NSF calls for funding boost in a bid to reverse decline in maths

The US National Science Foundation (NSF) needs to more than triple its commitment to mathematics over the coming years to reverse the subject's decline and meet the growing needs of other disciplines, Rita Colwell, the director of the agency, told the National Science Board last week."

- Taken from *Nature*
(Smaglik, 2000)

This is another cuick fix to all social policy problems – spend more money! It is shocking how predominant in government is the belief that increasing inputs to education will necessarily improve outcomes. The economic principle of the most cost effective allocation of scarce resources has thus far evaced policy makers in education.

This article also documents the ongoing math problem in the United States. In spite of huge increases in spending after Sputnik, the United States continues to rank poorly in international testing. Canada, and Newfoundland in particular, persists in following the loser. Our curriculum developers adhere to the documents produced by the professional body of teachers in the United States, the National Council of Teachers of Mathematics. The current emphasis in those documents puts problem solving as primary with skills acquisition as secondary. Yet cognitive psychology clearly states that you cannot problem solve effectively if you have to simultaneously use working memory to process basic skills. Human working memory is simply too small. An understanding of the underlying principles of basic skills must come first, then practice of those skills to automaticity must come second, and then, finally, use of those basic skills in problem-solving.

"Nothing worthwhile is lost, by taking the time to do it right."

- Abraham Lincoln (1809-1865), U.S.A.

Australia has a math problem too.

Here are some excerpts from the *Main Report* of the October 2003 publication entitled *Australia's Teachers: Australia's Future – Advancing Innovation, Science, Technology and Mathematics* (Australian Department of Education, Science and Training).

> Many areas of knowledge and skills are involved in
> creating a culture of innovation in addition to science-based
> research and development (R&D). Strategies are needed
> for the whole of schooling and all areas of the curriculum.
> But special emphasis is needed now on improving scientific
> and mathematical education and technological capability.
> (p. xvii)

The main report is 282 pages. There are two adjunct smaller documents, 82 and 48 pages, respectively. The only *documented* problem presented in this massive report is "a declining proportion of students who complete Year 12 studies in physics, chemistry and advanced mathematics" (p. xviii), with speculation that this decline works against the advancement of innovation in a technological world. Another problem is suggested but not documented –

> "some students who do not do well at school, including
> too many indigenous students, and [sic] may leave at the
> minimum permitted age with low attainments and poor
> motivation for continuing learning." (p. xviii)

There is data to support the low attainment and the dropouts, but none to support the hypothesis that these students would actually be better off with what would probably be a low-attainment high school graduation certificate. Our experience working with these students is that they require major intervention to assist them in achieving acceptable academic standards, intervention that is unlikely to be provided in an ordinary high school public education environment. What happens instead is that alternative education programs are provided that merely compromise normal academic standards. This serves no one well.

The Australian report proposes dealing with the situation by improving the teaching force. Presumably this is based on the assumption that to do so will significantly improve both participation and attainment. However, nowhere in the report is it actually documented that improving the qualifications of the teachers will increase participation or improve attainment. The report proposes massive, costly programs for the teaching profession based only upon professional speculation that significant improvements in the educational attainment of youth will result.

Like North America, Australia has a math problem. If the Australian government takes any heed of this October 2003 report, then Australia will have an even worse math problem in 2013. More students will have high school graduation certificates but fewer students will have useful employment skills. More students will have higher expectations for future opportunities but fewer skills to work towards those goals. More students will think of themselves as good problem solvers but they will be problem solving in a vacuum of skills. And the people will cry out against this travesty, and the government will respond by producing yet another useless or similarly damaging report.

Governments can generate reports detailing what might be but it is the student's responsibility to make it happen. And there is no easy way to do that. If it is easy, then something essential is missing. It is so convenient to expect a change in teacher training and/or the curriculum to provide the panacea, but until the learner becomes committed to the educational process being provided, there is absolutely nothing that the government or teachers or tutors or parents can do for that person. The current problem in Australia has nothing to do with a lack of innovation. It is a lack of confidence in the educational system, and that is a social problem not an educational one.

I return to Abraham Lincoln's quote: "Nothing worthwhile is lost, by taking the time to do it right." The flip side of this would be that nothing worthwhile is gained by taking the easy way out.

"Report finds maths education in need of urgent overhaul

Maths education is failing on every account and needs a fundamental multi-million pound overhaul, a government-backed review of the subject reported today."

- Taken from the *EducationGuardian* (Curtis, 2004)

This is a second section with an international flavor, this time from the United Kingdom.

Ironically, a somewhat similar report came out of the United Kingdom just four years ago. It was entitled *Skills for all: Proposals for a National Skills Agenda* (National Skills Task Force, 2000). This report documented major skills gaps in six areas including basic skills of literacy and numeracy, and mathematics skills. To address these shortages the report recommended a "new compulsory post-16 education and training system that ensured a sound base of basic skills, key skills and broader learning for all young people." The report proposed that the responsibility for this would rest with a "better trained" teaching force.

It would appear that this report fell short of the mark since four years later we read in the article cited on the left-hand page:

> The current system of GCSEs and A-levels is not meeting
> the needs of students, teachers, employers or universities,
> the report's author, Professor Adrian Smith, said today as
> he published the damning 186-page document, the result of
> a 15-month inquiry into the future of maths in schools.

Still, there is a belief that salvation rests with the teachers as we read on:

> Further incentives are necessary to recruit and retain more
> maths teachers.

The parallels between this report from the United Kingdom and the report cited from Australia (and earlier, from North America) are too predictable to bother repeating here. Similarly predictable will be the failure of this latest report from the United Kingdom to make any real difference in the educational outcomes of youth. The math problem is fundamentally a learning problem that crosses all academic curricula. Its solution must be attended to, not by governments or teachers, but by the learner him/ or her/ self.

"I simply expect people to be really, really good at what they do. With me, it's not a girl thing or a guy thing: it's a performance thing. Period. I give my players all the confidence they can handle and guess what? They usually rise to the top. But when they screw up, I hit 'em right between the eyes. Remember the old Westerns when they'd leave the back door at the bar open so they could escape if things turned bad? Well, I don't let my players leave the back door open. You've got to find a way out the front door in my program. None of this 'I was sick,' or "I didn't get what I needed.' I don't want to hear it. Be resourceful."

- Geno Auriemma, 47, women's basketball coach, University of Connecticut, U.S.A., taken from Where Pride Still Matters (Brooks, 2002)

In *The Little Book of Coaching*, Blanchard and Shula similarly speak about being conviction-driven. They maintain that "beliefs and convictions provide the boundaries and direction that people want and need in order to perform well," and that "inadequate beliefs are setups for inadequate performance" (2001, pp. 12-13).

Though many students with math difficulties may be labeled as below-average learners, we believe differently at the Mathematics Learning Centre. Most students who have trouble with math are very capable learners in other areas. These students, for some reason, haven't learned mathematics to the level they need, and so we try to understand that from the perspective of curricula issues, issues in cognitive psychology, or issues in intervention psychology. What educators often fail to realize is that the students perceive these other issues as pressing and sometimes more engulfing than skills development per se. In our experience, the associated impediments often pose the most difficult didactic challenges.

Our students must frequently be helped to deal with cognitive/emotional problems. Four such problems occur sufficiently often to mandate preparing staff to deal with them in a manner that promotes self-reliance and mutual respect. First, our students tend to perform more poorly in high school (grade point average of 70.4, 1990 to 1998 cohorts) than other university students (grade point average of 77.6, p < .001) (May, Rabinowitz and Mantyka, 2002, pp. 33-35). This substantial high school grade point difference reflects ability, skill, and motivation differences in the two populations. Thus, it is not surprising that many of our students must be taught study skills along with the realization that their actions largely determine what and whether they learn. Second, some of our students enter university believing that they are skilled at mathematics having never received a high school mathematics grade below 70. When informed that they must complete two or more semesters of remedial mathematics, they react with disbelief and hostility, often blaming us. Third and fourth, still other students enter university with either great mathematical anxiety or a conviction that they are incapable of learning mathematics. In an attempt to increase self-esteem and reduce mathematical anxiety, considerable effort is expended to insure that these students are successful at the beginning of their program with us.

In all these situations, we acknowledge that there is an explanatory factor to account for the educational outcome, but we do not absolve the student from having to take responsibility for turning it around. Beyond the initial acknowledgement of causality, we want the students to "be resourceful" and we will more than match their resourcefulness and commitment to achieving better educational outcomes.

With all this extraordinary support we give the students at the Mathematics Learning Centre, we believe we can demand the highest standards of performance from them, just as Geno Auriemma demands from her players. We are prepared to provide this support to our students but we expect them to take responsibility for their own learning. There are no part marks. Answers are right or wrong and a copying error makes an answer wrong. A pass usually requires more than 80% of the answers to be correct, and these pass standards are applied to each content section of a test. Every module test contains sections embodying pure skills and sections that are all word problems and *all* sections must be passed.

There is no back door out of our program either. This is because we know our program works. We have massive data sets to attest to the success our students have after they complete our program. Just as Don Shula can point to a perfect season with the Miami Dolphins in 1972, so too can we point to the graphs posted on our bulletin board and say to our students, "trust us – it works."

High School Grade Point Averages

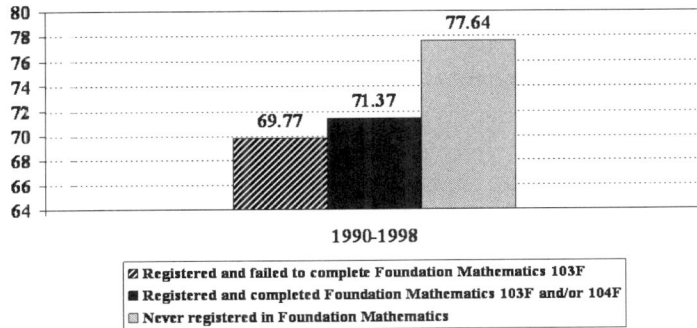

Bar chart for 1990-1998:
- 69.77
- 71.37
- 77.64

☒ Registered and failed to complete Foundation Mathematics 103F
■ Registered and completed Foundation Mathematics 103F and/or 104F
☐ Never registered in Foundation Mathematics

First-Year English Grade Point Averages

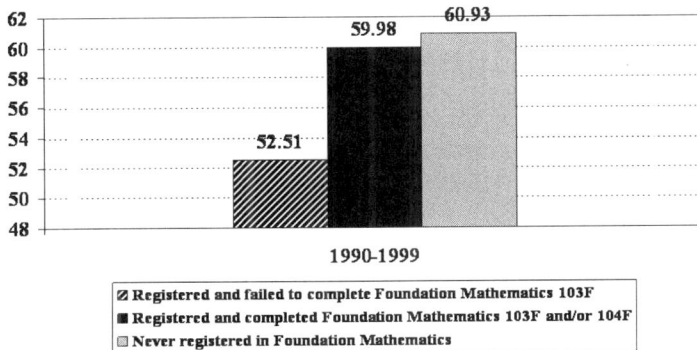

Bar chart for 1990-1999:
- 52.51
- 59.98
- 60.93

☒ Registered and failed to complete Foundation Mathematics 103F
■ Registered and completed Foundation Mathematics 103F and/or 104F
☐ Never registered in Foundation Mathematics

(May, Rabinowitz and Mantyka, 2002, pp. 50-51)

97

"HIGH SCHOOL DROPOUT RATE PLUMMETS

DOWN BY A THIRD IN 9 YEARS

Biggest declines reported in Atlantic provinces"

- Taken from the *National Post*
(Vallis, 2002)

In the introduction to this book, I mention two ways of meeting the social agenda of keeping youth in school longer. One is to improve programs and pedagogy; the other is to lower academic standards.

In Newfoundland in the early 1980s, high school graduation requirements were completely revamped in response to this social agenda. An extra year of high school was introduced, and a scheme of graduation credits was created. One consequence of the creation of these graduation credits was to allow for more "flexibility" in graduation standards, incorporating less stringent academic portfolios.

One has to ask how much better educated an individual is with a formal graduation certificate that is awarded based on compromised academic standards. One has to ask how much better off society is having expended more of its scarce financial resources to further educate that individual. One has to ask how much better off the individual is with an increased expectation of job opportunities available to him/her based on a graduation certificate of spurious worth to employers.

Everyone would have been better off if the social agenda had been more realistically addressed, but governments continue to yield to human weakness in looking for easy fixes to inherently tough problems. The current system lacks integrity and that serves no one well. We have shown at the Mathematics Learning Centre that individuals who are less academically inclined in mathematics can learn to do useful mathematics and that, in so doing, they learn to handle other academic subjects well, too. There is no need to patronize our youth by allowing them to graduate from high school having attained fewer useful skills in mathematics and language. These students must have access to more suitable learning paradigms that do not compromise academic standards. Only then will the desired outcomes from the laudable social agenda of the 1980s be attained.

"Bulldogs skipper vows to take title

BULLDOGS skipper Steve Price
says the team are [sic] still focused
on the NRL as the scandal-racked club
prepare to meet St. George Illawarra in
a trial on the Gold Coast tonight.

'It's been very hard this off-season,
we've trained probably better and
harder than we have over the past
five or six years,' Price said.

'I didn't train for six months not to win the
premiership this year.' "

- Taken from *The Herald* (2004, Feb. 28)

This excerpt about the Bulldogs was published amidst a series of much more spectacular articles about the Bulldogs with headlines such as "Lay off my boys – Folkes backs players as rape investigation intensifies" (Read, 2004). At the time of printing, six Bulldog players were being investigated by New South Wales police in an alleged sexual abuse of a 20-year old woman on Feb. 22 in the New South Wales coastal town of Coff's Harbour.

It is interesting to note the emphasis in the article about off-season preparation. The focus is on training and match fitness. I am reminded of something from *The Little Book of Coaching*. Ken Blanchard writes:

> Goal-setting is overrated! . . . . More important than setting
> the goals is the follow-up – attention to detail, demand
> for practice perfection, and all the things that separate
> the teams that win from those that don't. (Blanchard and
> Shula, 2001, pp. 42-43)

The Bulldogs' goal was to win the premiership title. The follow-up was to train hard for six months. The alleged sexual abuse could have been a major distraction from that training but neither the coaching staff nor the players wanted it that way. They remained absolutely focused on training as being the mechanism to take them "toward the vision of perfection, being the best their organization could be" (Blanchard and Shula, 2001, p. 43).

Against this backdrop, the following quote appeared a week later in *The Daily Telegraph* in an article entitled, "Season launch a night for healing":

The importance the game plays in people's lives is also a patent reminder of the responsibilities that are shared by everyone who is lucky enough to be a part of rugby league at the professional level.

More than ever it is the job of everyone in this room to keep sight of the strengths of rugby league, the skill, the excitement and the joy it brings to people week after week.

Having seen the Panthers turn around their fortunes in 2003, every team in this competition knows what can be done if you work hard. (Ritchie, 2004)

This is a recurrent theme in athletics. Over and over again we see an emphasis on skills, practice, hard work, and determination. So it must be with the pursuit of mathematics. To use mathematics in problem solving, there must first and foremost be an emphasis on skills, practice, hard work and determination. This is simply the essence of excellence.

"He had to have a well-trained army, and that meant drilling the soldiers over and over again, long after they thought they had mastered a technique, until it was so natural to them that they didn't have to think about it anymore."

- Taken from *Ender's Game* by Orson Scott Card (1994, p. 167)

"One who has mastered the art of living simply pursues his vision of excellence at whatever he does, leaving others to decide whether he is working or playing."

- James Michener
(Blanchard and Shula, 2001, p. 30)

I have a young friend who wants to be an opera singer. He trained very hard for six years, hoping to sing well enough to be admitted to a top-ranking school of music in the United States for further training.

For six years his training to prepare for auditions for post-graduate study entailed all of the following:

(a)   Development and daily conditioning of the muscles that are used to produce singing tones and breathing capacity to sustain those tones continuously and melodically. At a minimum this means singing intently every day for at least two hours.
(b)   Regular and extraordinary singing lessons.
(c)   Piano lessons (compulsory, to better understand the connection between the vocalist and the accompanist).
(d)   The study of the history of music, music composition, orchestration, and other related academic areas.
(e)   The study of foreign languages and their elocution to facilitate singing opera in German, Italian, French, etc.
(f)   Singing in choirs and other vocal groups.
(g)   Memorizing specific pieces of music for concert performance.
(h)   Practicing specific pieces of music for concert performance.

All of this, and in particular, repeated practice of the performance pieces, frees the vocalist from having to attend to the musical rudiments in performance. Instead, the vocalist can enter fully into the emotional expression of the music, can attend to the fine-tuning of the most difficult sections, and can concentrate on covering up any mistakes made.

As much as this young man loves to perform, most of what went into his preparation for a possible opportunity at further training was drudgery. But he understood that there is tough competition to get into the best U.S. schools and this is what he had to do just to have a chance.

Interestingly this young man also does very well in mathematics courses even though he feels uncomfortable with the subject and does not enjoy doing it at all. He succeeds at mathematics by using a structure for learning math similar to what he has been trained to do to learn how to sing well. This means he attends all his math classes, he practices what he has learned each day for one to three hours, he gets help with anything troublesome in the very next class, he commits to memory recurring rules, facts, and algorithms via thoughtful repetition in practice problems, and he continually uses mathematics to make sure he retains his skills. That is, he studies mathematics with the same discipline, intensity and regularity that he does to train as a vocalist.

My young friend does not need mathematics for his career. Still he studies it seriously because it is in his character to be that way about whatever he chooses to learn. If he spends any time at it, then he wants to be sure he will gain something from the experience. He knows that without commitment and regular practice, nothing useful will result.

"If I don't practice one day, I know it; two days, the critics know it; three days, the public knows it."

- Jascha Heifetz, Violinist, 1901-1987
(Heifetz, 2006)

"Winning is not a sometime thing; it's an all-time thing. Winning is a habit. Unfortunately, so is losing."

- Vince Lombardi

I have been taking dancing lessons in many different types of dance for more than forty years. I started lessons when I was twelve years old. In those early days I learned simply by imitating the movements of the instructor, and I remembered sequences of steps by counting beats in the rhythm of the music.

As an adult learner, I was less content with learning purely by imitation, and I had little success remembering by merely counting beats in the music. In addition to a demonstration of the movements, I needed some kind of verbal description of what was intended. For example, this is a verbal string I created to describe what I saw Evelyne Benais of El Viento Flamenco demonstrate in an *Escabeya*:

1	on L, planta, tacon
2	golpe R R
3	golpe L
& 4	brush across R
5	brush back
6	place R (no weight, with circular leg motion)

then repeat on R
then as on left 1, 2, 2, &4, &5,

&6	planta R behind, golpe L
4, 5, 6	repeat twice

Repeat all to R

I needed this verbal string to cue my movements so that I could repeat it many times, accurately, at increasing speed. In this manner, I would eventually be able to do the sequence of steps automatically; that is, without deliberately recalling each word in the verbal string from memory.

If you have read most of this book (or, specifically, "We Learn by Doing", p. 14), you will recognize the description in the first paragraph as implicit learning, and the description in the second paragraph as explicit learning. What was being learned did not change – I did. Furthermore, it was me who made the adjustment to the learning experience and not the instructor.

This is what many students who have under-achieved in mathematics have to do. It is not necessarily true that they have trouble learning math, but they may have trouble learning the way they are being taught. So they have to reshape the learning experience to something that works better for them but that will achieve the same outcome.

As Lombardi says in his quote, winning and losing are both habit-forming. That is because it is largely a matter of attitude. It makes sense to use what works well for you as much as you can. Sometimes this requires an injection of creativity but it can be done. This is the stuff winners are made of.

Evelyne Benais
El Viento Flamenco

"Fate Offers Synchronicity

**The car accident was pivotal.
It allowed me to dream different dreams,
and live them. I have always appreciated
that fate gave me the opportunity to
thank the man whose car hit mine."**

- By Shari Graydon, taken from
The Globe and Mail
(Graydon, 2002)

My best friend in high school and my first four years of university is a visual artist. Her name is Agnes Ruest. By our early teen years Agnes knew she wanted to be an artist and I knew I wanted to be a mathematician. We were an odd pair because of our dissimilar preferences for profession. What we had in common were fathers who would not let us date and so we were social misfits with most girls our age.

My friendship with Agnes profoundly affected my professional life as a mathematician. There is a common belief that the analytical, scientific mind cannot be at one with the passionate, creative mind. Agnes and I have lived a different experience. Because of our friendship in high school, Agnes took the same science and mathematics courses as I did. We did our practice problems shoulder to shoulder in class but separately in thought. We each did more practice problems at home on our own, and both of us did very well in those subjects. Agnes did a degree in Fine Arts at university and further teacher training for the visually impaired. Now she has full-time teaching responsibilities in high school mathematics and science for the visually impaired, and she is a very successful visual artist.

For my part, the self-awareness and confidence I gained through my connection with art through Agnes significantly altered my career path as a professional mathematician. I have allowed myself to venture outside the analytical box of career academia in mathematics to express my passion for life in my approach to doing mathematics. In turn, this has influenced me to encourage hundreds of students to view mathematics initially through their natural creative perspective instead of a more alien analytical framework, and then reconcile the two.

There is an excerpt from Shari Graydon's article which eloquently encapsulates the process of reconciling seemingly dissonant attitudes to circumstances:

I reflected on the drawing I'd been devoted to as a child, the acting and dancing I'd immersed myself in at university, and the journal-writing I'd been wedded to while traveling in Europe. My father called me a "massive communicator" because I told jokes, expressed my emotions, wrote poetry – and then actually shared it with the people who'd inspired it.

I thought about my experience of giving people the gift of massage. As rewarding as it was to ease sore muscles and re-acquaint people with neglected parts of their bodies, it required me to do all my communicating through my hands. The monologue varied little from one day to another, the arc of the story was generally the same, and I usually tried to avoid making my audience laugh or cry. On a good day, they'd fall asleep during the performance.

I decided I didn't want to go to Toronto or Colorado; I wanted to use words to make a difference to the world. I wanted to write and to speak. I wanted to move people. To make them laugh, and think. (Graydon, 2002)

We express ourselves in everything we do – in our friendships, in our leisure activities, in our studies, and in our work. To be whole people and to make the most of our lives, it is essential to entrust ourselves to each experience to achieve one goal. We must first of all know what that goal is. Then we must always be open to scrutiny of our experiences to know if they fit. If they do not, we must try to understand why that is so. It is impossible to study a subject like mathematics without passion; the level of commitment required to learn it well demands that. The passion need not derive from a love of the discipline. Rather, the source may be a determination to achieve some end that requires it, or to prove to yourself that you can overcome this obstacle.

Shari Graydon found an opportunity in the car accident to refocus her life on something that allowed her to express her passion for life in her work. I am offering the reader this chance without enduring the inconvenience and discomfort of a comparable setback. This is a great opportunity for the reader – declare your calling and find the outlet for your passion now. While you are doing so, you can discover if there is any genuine synchronicity with mathematics in your future.

"Adaptation is not allowing yourself to give in to circumstances; it's allowing those circumstances to give you success."

(Blanchard and Shula, 2001, p. 56)

I once attended a workshop in African drumming and dancing because I was interested in dancing. I had never even touched a percussion instrument so I certainly did not think the drumming section of the workshop would be of any benefit to me. However, when the workshop began I was horrified to discover that we were all going to have to do some drumming first. I was mortified.

I survived the drumming. In fact, I enjoyed it more than the dancing because the rhythms were fascinating. They were unlike the rhythms in Western music in so many ways. There was no counting, the phrasing was irregular, the cues to begin were not specific, and we were not taught sequences or patterns to remember. What we were exposed to were rhythm packets and musical cues to make transitions from one rhythm packet to another. We were expected to be able to duplicate specific rhythm packets simply by having imitated them and then repeated them many times.

The initial focus in the workshop on percussion was to familiarize us with the rhythms. This is because the dance was basically a physical response to the rhythm as opposed to a choreographed narrative. If we had a good sense of the rhythm, the dance would flow from that.

I was so taken with this exposure to African drumming and dancing that I traveled to Ghana with two other Canadians and the master drummer, Kwasi Dunyo, who gave the workshop. For three and a half weeks in Ghana, we lived with Kwasi's family and experienced African music. Our lessons in percussion were essentially jam sessions in Kwasi's home village where local men would gather with their drums under a tree and just begin to play. We were expected to develop a sense of the rhythms and join in – no counting, no verbal instructions, just beating on the drum. As an adult I had never had to learn anything this way before and I was floundering. Nevertheless, by the end of the three and a half weeks I could hold my own and even make it through as the lead drummer. But if you had asked me to describe how to do it at the time, I would not have been able to do so even vaguely. I was simply responding to a complete

117

immersion in the experience of the music, specifically the various rhythms.

As an adult learner, this was the first time I had been forced to learn something completely implicitly. It was so alien to me. I felt very uncomfortable learning this way. It gave me a first-hand experience of what it is like for many of the students at the Mathematics Learning Centre who, as natural implicit learners, have been forced to learn mathematics explicitly. But it also gave me a vision of how to bridge the gap.

When I got back to Canada I was anxious to record what I had learned. I contacted Professor Russell Hartenberger, Faculty of Music, University of Toronto, who is very knowledgeable about percussion and African music. I asked him for a reference in which African rhythms were recorded. I was stunned when he told me there were none; written records detailing these rhythms did not exist. I was going to have to create a vocabulary to record what I had learned. So I did (Mantyka, 2006).

This is one of the challenges that faces many natural implicit learners who are trying to learn mathematics. If they have trouble learning some bit of mathematics explicitly, they have to simply imitate the pattern of the technique as many times as it takes for them to be able to successfully duplicate it without error. This allows the learner to find a rhythm in the execution of the various steps (and there always is one). Then they have to create for themselves some cues for memory retrieval of that rhythm. The more complicated the technique, the more practice is required and the more complex the trigger.

This is not an easy or efficient process for learning but it can be done with determination and creative energy, characteristics common to many capable implicit learners. This is probably what Ken Blanchard had in mind when he wrote what I quoted at the start of this section, but to me it is a two-tiered hierarchical application of the one principle.

My youngest daughter, Janys, learned calculus this way, and it was her intuition into the process that gave me the idea for this section of the book. As she was doing practice problems from her calculus textbook, she looked up and said to me, "the difficulty is, each subsequent problem is so different from the previous one that I can't sense any rhythm in the techniques" (J. May, personal communication, Feb. 20, 2002). This then is another problem with the modern textbooks that emphasize problem solving over skills – not enough problems of a similar type to allow the implicit learner to develop the sense of rhythm of the technique that comes with repeated practice. There is neither enough practice problems for the explicit learner to overlearn the skill to automaticity, nor is there enough practice problems for the implicit learner to develop a sense of rhythm. Who do these modern textbooks serve? They serve the teachers who are bored with the old curriculum and simply want a change. Textbooks written for the teachers and not the students – now that's a novel educational concept.

"Many people who have never had occasion to learn what mathematics is, confuse it with arithmetic and consider it a dry and arid science. In actual fact it is the science which demands the utmost imagination. One of the foremost mathematicians of our century says very justly that it is impossible to be a mathematician without also being a poet in spirit. It goes without saying that to understand the truth of this statement one must repudiate the old prejudice by which poets are supposed to fabricate what does not exist, and that imagination is the same as "making things up." It seems to me that the poet must see what others do not see, must see more deeply than other people. And the mathematician must do the same."

- Sofya Kovalevskaya (1850-1891)
(Kovalevskaya, 1978, p. 35)

I would characterize this entire book as an example of Sofya Kovalevskaya's vision of mathematics as declared in the quotation just cited. This is a book about mathematics but it is also a book about music, football, psychology, behavior, coaching, and practice. This book is the product of my being given a variety of opportunities that have allowed me to see more deeply than other people. This book is not something which I made up. It is a summary of connections I have made that others have not. Interestingly, mathematics is often called the science of pattern recognition and is promoted as a discipline that requires one to make connections. Yet, most students compartmentalize their knowledge of mathematics and that is the antithesis of using it to make connections.

In this book I have tried to make connections to mathematics of as many disparate pursuits and activities as I could. To maintain a working interest in this subject, the reader must do the same. To kindle the kind of commitment to mathematics that you need to develop a working knowledge of it, you must make a connection between mathematics and something you feel passionate about. Only then will be you able to sustain your involvement with mathematics to get in all the boring, repetitious practice that comes with being able to do useful mathematics. Just as this book is a manifestation of the poet in me, you too must find the poet in yourself.

"When you finally decide how successful you really want to be, you've got to set priorities. Then, each and every day, you've got to take care of the top ones. The lower ones may fall behind, but you can't let the top ones slip. In 25 years as a head coach and an assistant, I think I might have missed one practice. Why? Because practice is my top priority. A day doesn't go by when I don't accomplish something in my family life or in my profession, because those two things are my top priorities."

- Dan Gable, 53, Olympic champion and former wrestling coach, University of Iowa, U.S.A., taken from Where Pride Still Matters (Brooks, 2002)

There is a fine line between arrogance and confidence. There is a delicate balance to be achieved in having the confidence to take risks without ego involvement to insulate you from possible failure.

For me achievement of this balance requires solitude and silence to move deep within myself to remind me of my priorities, my goals, my values. My external trappings are there to provide me with an initial interest as I try to move my relationships beyond those trappings to a greater place within. This place within is characterized by genuine caring and the sensitivity, generosity and compassion that come with it. This place within exists in all of us, but for some it is buried beneath the residue of betrayal, deceit, disappointment and failure. For most of us, we only get glimpses of our interiority occasionally and it sustains us through certain moments of our lives. For these moments we must be truly grateful because it is these experiences that have the power to transform us and truly change the quality of our lives.

For many under-achievers in mathematics, it is this quality of experience that is required to enable the individual to improve their performance. Nothing less will do. It is a tall order for both the teacher and the learner because it is about our humanity and not just our intellect.

I have frequently been described as a mathematician with a heart. This is more a statement about the common perception of mathematicians than it is a statement about me. Mathematicians do have a distasteful public image (recall the quotation in an earlier section of this book beginning "Math teachers are badly dressed . . .", p. 12) and this is not helped by the media representation of us. But while our sense of fashion is unlikely to be adopted by today's youth, it is not true that our strong analytical skills negate our emotional attachment to our students. We may hide behind our nerdy glasses and present our backs to classrooms filled with frustrated, hostile students, but it is not because we do not care about our students. It is because we do care and we, too, are frustrated by the current situation.

There is so much that can easily be done to rectify the current dismal situation in the teaching/learning environment of mathematics. The gains would be greater if we could work together towards this goal, but it is not impossible to pursue it individually and/or in small groups. This book has been designed to give everyone hints on how to go about it. I hope you have found something in here to get you started. The rest is up to you.

"A 'demanding coach' is redundant. A coach sets the standards. But you need balance. You have to laugh with them about the toughness of the game, the human condition; it's got a lot of failure in it, just like baseball does. You can demand a lot from people if you care about them. If they perceive themselves as objects of your ego, you can't teach them. If they are going to be happy with you and produce, they have to know you care."

- Wayne Graham, 64, baseball coach,
Rice University, U.S.A.,
taken from Where Pride Still Matters
(Brooks, 2002)

IT WAS A LITTLE KNOWN BUT SOMETIMES USEFUL FACT THAT DRACULA HAD A SEVERE CASE OF MATH ANXIETY.

Note: From *Not Strictly by the Numbers* by H. Blair and B. Knauff, 1991, Burlington, NC: Carolina Biological Supply Company. ©1991 Carolina Biological Supply Company, Burlington, NC. Used by permission.

BIBLIOGRAPHY

References marked with an asterisk indicate studies included in the meta-analysis.

A Tale of Two Approaches to Policy Development: Evidence-Based Versus Knee-Jerk Anecdotal, A Case Study of the Mathematics Learning Centre at Memorial University, Government of Newfoundland and Labrador, January 2002.

AFC Championship: Patriots 24, Steelers 17. (2002, January 28). *The Boston Globe*, p. A1.

Ashcraft, M.H., & Kirk, E.P. (2001, June). The Relationships Among Working Memory, Math Anxiety, and Performance. *Journal of Experimental Psychology: General, 130*(2), 224-237.

Australian Department of Education, Science and Training. (2003). *Australia's Teachers: Australia's Future – Advancing Innovation, Science, Technology and Mathematics* (Vols. 1-3). Australia: Commonwealth of Australia.

Bernhardt, D. (2002, Apr. 13). Math specialists lacking in Saskatoon classrooms: board. *The StarPhoenix*, p. A7.

Berry, D.C., & Dienes, Z. (1993). *Implicit Learning: Theoretical and Empirical Issues*. East Sussex, UK: Lawrence Erlbaum.

Blair, H., & Knauff, B. (1991). *Not Strictly by the Numbers*. Burlington, NC: Carolina Biological Supply Company.

Blair, J. (2002, Feb. 28). Players show no sign of being lame ducks. *The Globe and Mail*, p. S3.

Blanchard, K., & Shula, D. (2001). *The Little Book of Coaching: Motivating People to be Winners*. New York, NY: Harper Collins Publishers.

Brooks, D. (2002, June). Where pride still matters. *Men's Health*, pp. 80-83.

Bruckheimer, J., & Oman, C. (Producers), and Yakin, B. (Director). (1989). *Remember the Titans* [Motion Picture]. Burbank, CA: Disney Enterprises.

Bulldogs Skipper Vows to Take Title. (2004, Feb. 28). *The Herald,* pp. 74-75.

Campbell, J.I.D., & Xue, Q. (2001, June). Cognitive Arithmetic Across Cultures. *Journal of Experimental Psychology: General, 33*(2).

Canfield, J., & Hansen, M.V. (1993). *Chicken Soup for THE SOUL*, p. 29. Deerfield Beach, FL: Health Communications.

Card, O.S. (1994). *Ender's Game*. New York, NY: Tom Doherty Associates, LLC.

*Collins, M., & Harding, L. (2004, Jan. 22). Math Learning Centre applies new techniques to help students learn math. *The MUSE*, p. 9. Also retrievable from http://www.mun.ca/mlc.

Cox, K. (2002, March 23). Low marks sending N.S. pupils back to mental arithmetic. *The Globe and Mail*, p. A6.

Curtis, P. (Feb. 24, 2004). Report finds math education in need of urgent overhaul. *EducationGuardian.co.uk*. Retrieved Feb. 24, 2004, from http://education.guardian.co.uk/schools/story/0,5500,1154932,00.html.

*Dalziel, A. (2002, Mar. 8). Bridging the gap. *Gazette*, p. 9. Also retrievable from http://www.mun.ca/mlc.

Dubner, S.J., & Levitt, S.D. (2006, May 7). A Star is Made. Retrieved May 24. 2006 from http://www.nytimes.com/2006/05/07magazine/07wwln_freak.html.

El Viento Flamenco (2006). Photos. Retrieved July 17, 2006, from http://www.elvientoflamenco.com/gallery/pics-evf.htm.

Etchegary, V. (1999, Fall). The members of Newfoundland's most popular foursome are this year's alumni of the year. *Luminus, 25*(1). Retrieved Mar. 14, 2004, from http://www.mun.ca/munalum/luminus11/07.html.

*Gaskill, H.S. (2000, Sept. 3). Math problem can't be ignored. *The Telegram*, p. 10.

*Gaskill, H.S. (2000, Sept. 2). Math problems, solutions. *The Telegram*, p. 10.

Graydon, S. (2002, Mar. 1). Fate offers synchronicity. *The Globe and Mail*, p. A18. Also retrievable from http://www.mun.ca/mlc.

Grieve, K. (2002, Mar. 1). Lives lived: Eileen Nash. *The Globe and Mail*, p. A18.

*Hanushek, E.A., & Kimko, D.D. (2000, Dec.). Schooling, Labor-Force Quality, and the Growth of Nations. *The American Economic Review, 90*(5), 1185-1208.

Haverty, L.A. (1999). *The Importance of Basic Number Knowledge to Advanced Mathematical Problem Solving.* Unpublished doctoral dissertation, Carnegie Mellon University, Pittsburgh, PA.

Heavin, G., Findley, C., & Thomas, A. (2005). *Curves Member Guide.* Waco, TX: Curves International, Inc.

Heifetz, J. (2006). Quotes. Retrieved July 13, 2006, from http://www.jaschaheifetz.com.

*Henry, W.A. (III). (1994). *In Defense of Elitism.* New York, NY: Doubleday, a division of Bantam Doubleday Dell Publishing Group, Inc.

Holley, M. (2002, January 28). Bledsoe's comeback is a real tear-jerker. *The Boston Globe*, pp. D1, D8.

Holt, J. (1993). We Learn by Doing. *Chicken Soup for the Soul*, p. 132. Deerfield Beach, FL: Health Communications.

Hutchinson, C. (2006, June 17). Armstrong Leading the Pack -- to the Bar. *The Star Phoenix*, p. B3.

In the news . . . (2001, Jan. 22). Social Studies section. *The Globe and Mail*, p. A16.

JOHNSRUDE. (2002, Apr. 19). OKAY, LISTEN UP! [Cartoon]. *The StarPhoenix*, p. A15.

Johnston, W. (1999). *Baltimore's Mansion.* Canada: Alfred A. Knopf.

Kovalevskaya, S. (1978). *A Russian Childhood.* New York, NY: Springer-Verlag.

Lang, M. (2002, Apr. 8). Poor math skills stump experts. *The StarPhoenix*, p. A3.

Li, L. (2003). *Analysis of first year math courses*, Tables I.1.2 and I.5.3. Internal Report, Department of Mathematics and Statistics. St.

John's, NL: Memorial University. Retrieved Sept. 2, 2006 from http://www.mun.ca/mlc.

Logan, G.D. (1988). Towards an instance theory of automatization. *Psychological Review, 95,* 492-527.

MacMullan, J. (2002, January 28). This title is shrouded in secrecy. *The Boston Globe,* p. D8.

Mandel, C. (2002, Feb. 27). U2 and Denise who? *The Globe and Mail,* p. R5.

Mantyka, S. (2006). Answers to *The Math Plague* Matching Quiz. http://www.mun.ca/mlc.

Mantyka, S. (2006). Implicit Learning: The Gahu Rhythm of West Africa. http://www.mun.ca/mlc.

Mantyka, S. (2004, March 3). Resourceful Adaptation. Invited address to *Newcastle Enterprising WOMEN,* Juicy Beans Café, Wheeler Place, The Civic, Newcastle, NSW, Australia.

Mathematics: Primary. (August, 2000). Curriculum Guide. Government of Newfoundland and Labrador: Division of Program Development, Department of Education.

May, S. (2000, Sept. 17). Math isn't easy without first learning the basics. *The Telegram,* p. 13. Retrieved September 21, 2006 from http://www.mun.ca/mlc.

May, S. (2003, Sept. 24). Participation of Current Matriculants in Mathematics Courses at Memorial University 1998-2002. Internal Report, Office of the Dean of Science, pp. 3-4. St. John's, NL: Memorial University. Retrieved September 21, 2006 from http://www.mun.ca/mlc.

May, S., Rabinowitz, F.M., Hart, S., & Larson, K. (1995). Mathdrill (Version 1.0) [Computer software]. St. John's, NL: Mathematics Learning Centre, Memorial University.

May, S., Rabinowitz, F.M., & Mantyka, D. (2001). Teaching the Rules of Exponents: A Resource-Based Approach. In F. Glandfield (Ed.), *Mathematical Understanding: Four Perspectives* (pp. 63-77). Saskatoon, SK: University of Saskatchewan Press.

May, S., Rabinowitz, F.M., & Mantyka, D. (2002). Teaching Remedial Mathematics at the University: Rationale, Principles, Procedures, and Outcomes. Retrieved June 23, 2006 from http://www.mun.ca/mlc.

McDonough, W. (2003, January 28). A dream sequence for player who never quit. *The Boston Globe*, p. D8.

National Council of Teachers of Mathematics. (1989). *Curriculum and Evaluation Standards for School Mathematics*. USA: National Council of Teachers of Mathematics.

National Skills Task Force. (June 2000). *Skills For All: Proposals for a National Skills Agenda* (Ref. SKT 28). Sudbury, Suffolk, UK: Final Report of the National Skills Task Force. Retrieved June 2000 from http://www.dfee.gov.uk/skillsforce.

Next stop, Super Bowl. (2002, January 28). *The Boston Globe*, p. D9.

Norman, D.A., & Shallice, T. (1986). Attention to Action: Willed and Automatic Control of Behavior. In R.J. Davidson, G.E. Schwartz, & D. Shapiro (Eds.), *Consciousness and Self-Regulation: Advances in Research* (Vol. 4, pp. 1-18). New York, NY: Plenum.

Palmer, M. (2002, Feb. 4). Basic skills remain key to the game. *The Globe and Mail*, p. 56.

Pearce, T. (2002, Mar. 2). A few gems amid the trashy grammy divas. *The Globe and Mail*, p. 12.

Read, B. (2004, Feb. 28). Lay off my boys. *The Herald*, p. 96.

Ritchie, D. (2004, Mar. 4). Season launch a night for healing. *The Daily Telegraph* (Sydney, NSW, Australia), pp. 74-75.

Rowling, J.K. (2000). *HARRY POTTER and the Philosopher's Stone*, p. 126. Vancouver, BC: Raincoast Books.

Salisbury, D.F. (1990). Cognitive Psychology and Its Implications for Designing Drill and Practice Programs for Computers. *Journal of Computer-Based Instruction*, *17*, 23-30.

Scattergood, T. (1884). Motto Calendars. Pottsdown, PA: email Mottocal@aol.com.

Schorr, L.B. (1989). *Within Our Reach: Breaking the Cycle of Disadvantage*. New York, NY: Anchor Books.

*Shames, L. (1991). *The Hunger for More: Searching for Values in an Age of Greed*. New York, NY: First Vintage Books, a division of Random House , Inc.

Shaughnessy, D. (2002, February 4). CHAMPIONS! Patriots win Super Bowl, 20-17, on Vinatieri's last-second kick. *The Boston Globe*, p. A1.

Smaglik, P. (2000, Oct. 26). NSF calls for funding boost in a bid to reverse decline in maths. *Nature*, Vol. 407, p. 931.

Smith, A. (2004). *Making Mathematics Count: The Report of Professor Adrian Smith's Inquiry into Post-14 Mathematics Education.*

Retrieved Feb. 27, 2004, from http://image.guardian.co.uk/sys-files/
education/document/2004/02/24/Maths_ Review_Text.pdf.

TSN SPORTSCENTER. (2002, Feb. 1).

Teachers: A Year 2006 Daily Calendar. (2005). Cypress, CA: Avalanche
Publishing, Inc.

Vallis, M. (2002, Jan. 24). HIGH SCHOOL DROPOUT RATE
PLUMMETS. *National Post*, p. A4.

Watson, C. (2002, Nov. 7). Holy roller role for *Jets* rocker. *The Herald*
(Newcastle, NSW, Australia), p. 31.

Weight Watchers Offer New Year's Diet Tips. (2002, Jan. 5).
applesforhealth.com 2(32). Retrieved June 8, 2004, from
http://www.applesforhealth.com/ wwdiettip2.html.

Wortzman, R., Harcourt, L., Kelly, B., Morrow, P., Charles, R.I.,
Brummett, D.C., *et al.* (1996). *Quest 2000: Exploring
Mathematics*. Teacher's Guide and Journal. Don Mills, ON:
Addison-Wesley Publishers Ltd.

Sherry Mantyka is available as a motivational speaker for issues related to under-achievement in mathematics. She is also available to speak about teaching and learning issues in mathematics for a more general audience.

As a speaker Sherry is passionate, dynamic and engaging.

For more information, please contact The Mathematics Learning Centre at:

Telephone: 709-737-3308
Fax: 709-737-2351
Email: mlc@mun.ca
Website: http://www.mun.ca/mlc

- -

To order this book, please contact the publisher at:

Telephone: (709) 579-5879
Email: maytcc.smay@gmail.com